Fundamentals of IoT and Wearable Technology Design

IEEE Press
445 Hoes Lane
Piscataway, NJ 08854

Technical Reviewers

Khalid Abed, PhD, Professor, Department of Electrical & Computer Engineering and Computer Science, Jackson State University, MS, USA

Tareef Al-Mahdawi, PhD, Sr. Principal Scientist II. BAE Systems Fast Labs, USA

John B. Anderson, PhD, Professor in Digital Communication Emeritus Electrical and Information Technology, Lund University, Sweden

Ali Hammodi, PhD, Project Manager, Amphenol Inc, USA

Fundamentals of IoT and Wearable Technology Design

Haider Raad

Department of Physics, Engineering Physics Program
Xavier University

Published by John Wiley & Sons, Inc., Hoboken, New Jersey.

Published simultaneously in Canada.

For general information on our other products and services or for technical support, please contact our Customer Care Department within the United States at (800) 762-2974, outside the United States at (317) 572-3993 or fax (317) 572-4002.

Wiley also publishes its books in a variety of electronic formats. Some content that appears in print may not be available in electronic formats. For more information about Wiley products, visit our web site at www.wiley.com.

Library of Congress Cataloging-in-Publication Data:

Names: Raad, Haider Khaleel, author.
Title: Fundamentals of IoT and wearable technology design / Haider Raad.
Description: First edition. | Hoboken, New Jersey : John Wiley & Sons, Inc., [2021] | Includes bibliographical references and index.
Identifiers: LCCN 2020042822 (print) | LCCN 2020042823 (ebook) | ISBN 9781119617532 (hardback) | ISBN 9781119617549 (adobe pdf) | ISBN 9781119617556 (epub)
Subjects: LCSH: Wearable technology. | Internet of things.
Classification: LCC TK7882.W435 R33 2021 (print) | LCC TK7882.W435 (ebook) | DDC 621.381–dc23
LC record available at https://lccn.loc.gov/2020042822
LC ebook record available at https://lccn.loc.gov/2020042823

Cover Design: Wiley
Cover Image: © yosart / Shutterstock

Set in 9.5/12.5pt STIXTwoText by SPi Global, Pondicherry, India

To my family with love

Contents

About the Author

Haider currently serves as the director of the Engineering Physics program and the Wearable Electronics Research Center (XWERC) at Xavier University in Cincinnati, OH, USA. He was previously affiliated with California State University and the University of Arkansas, Little Rock between 2008 and 2015.

Haider received the Ph.D. and M.S. degrees in Systems Engineering, specializing in RF Telecommunication and Wireless Systems from the University of Arkansas at Little Rock (UALR), and the M.S. degree in Electrical and Computer Engineering from New York Institute of Technology (NYIT).

Haider teaches several courses such as Electronic Circuits, Microprocessors and Digital Systems, Communication Systems, Antenna Engineering, and Control Theory. He has given over 50 lectures at universities around the world and is a frequent speaker at international conferences. Professor Raad is also connected to the industry through his engineering consulting firm.

Haider has published five books in the fields of Wearable Technology, Telemedicine, and Wireless Systems. He has also published over a hundred peer-reviewed journal and conference papers on research fields of his interest which include: Flexible and Wearable Wireless Systems, Telemedicine and Wireless Body Area Networks, IoT, Metamaterials, and Biomedical Electronics. He is also the recipient of the 2019 Outstanding Teaching Award, the 19th International Wearable Technology Conference Best Paper Award in 2017, the E-Telemed Conference Best Paper Award in 2016, Sonoma State University's Research Fellowship Award in 2015, and AAMI/TEAMS Academic Excellence Award in 2012.

Haider loves spending quality time with his family. In his free time, he enjoys composing world, rock, and smooth jazz music. He also enjoys street photography and artistic activities.

Preface

Everything will be connected. This is one of the rules that will govern the future. And contrary to popular belief, the impact of Internet of Things and Wearable Technology will be much greater than a smart light bulb or a fitness tracker. Connecting everything will dramatically reshape our world in ways we can barely imagine.

Locating a wandering Alzheimer's patient by sensors embedded within lighting poles in a smart city, or detecting if a driver is having a heart attack by analyzing vital signs and facial expressions by a system integrated within a vehicle's dashboard, are just a couple of scenarios these technologies will be capable of doing. We will also witness the fantasy of fully automated smart cities and driverless vehicles work in coordination with one another fairly soon.

Today, IoT and Wearable Technology are recognized as two of the fastest-growing technologies and hottest research topics in academia and research and development centers. Wearable devices, which are characterized by being lightweight, energy-efficient, ergonomic, and potentially reconfigurable are expected to substantially expand the applications of modern consumer electronics. Similarly, there has been a massive interest in smart objects that can be connected to the Internet allowing remote access, processing, and control, which enable innovative services and applications. Such objects are utilized in smart homes, healthcare, power grids, transportation, and numerous other industrial applications.

Although IoT and wearable devices are electronic systems by definition, the study of these interrelated technologies is multidisciplinary and borrows concepts from electrical, mechanical, biomedical, computer, and industrial engineering, in addition to computational sciences. Having worked in this field for almost 12 years in both academic and industrial capacities, I feel the need to compile a comprehensive technical resource that academically tackles the various design aspects of these technologies.

The aim of this book is to provide an extensive guide to the design and prototyping of IoT and Wearable Technology devices, in addition to introducing their detailed architecture and practical design considerations. The book also offers a detailed and systematic design and prototyping processes of typical use cases, covering all practical features. It should be noted that this book attempts to address the design and prototyping aspects of these technologies from an engineering/technical perspective rather than from a maker/hobbyist perspective.

The intended audience of the book encompasses both undergraduate and graduate students working on projects related to IoT and Wearable Technology. The book also serves as an extensive resource for research and development scientists, university professors, industry professionals, and practicing technologists.

It is worth noting that familiarity with fundamental computer programming, mathematics, electricity laws and properties, digital and information theories, and basic networking and computer architecture is required to understand the topics covered in this book.

Chapter 1 of this book helps the reader understand what IoT and wearables exactly are and examine their characteristics. The chapter also provides an overview of the history and beginnings of IoT and Wearable Technology and aims at demystifying the differences between the two.

Chapter 2 covers the applications of IoT and wearables in various fields. It also provides an insight on the roles these applications could play in practice and discusses the challenges and key success factors for their adoption.

Various architectures used in IoT and wearable devices along with important architecture concepts will be discussed in Chapter 3. Further, simplified and versatile architectures are proposed to help the reader articulate the key functions and elements of IoT and wearable devices.

Chapter 4 highlights the capabilities, characteristics, and functionality of sensors and actuators with an understanding of their limitations and their role in IoT and wearable systems. Criteria for selecting microprocessors and communication modules will be discussed next. Additionally, deciding on a suitable energy source with a matching application-specific power management design is discussed. Finally, the reader will gain an understanding on how to bring these foundational elements together to realize a smart devices that makes most IoT and wearable use cases possible.

Chapter 5 takes a look at the characteristics and basics of the communication protocols that IoT and wearables employ for their data exchange, along with a dive into some of the most common technologies being deployed today.

Chapter 6 discusses the development process and design considerations that developers must follow to guarantee a successful launch of IoT and wearable products.

Chapter 7 provides an overview of cloud topologies and platforms, and an architectural synopsis of OpenStack cloud. Next, Edge topologies and computing technologies will be presented. It will be shown that the maximum value from an IoT or wearable technology project can only be gained from an optimal combination of cloud and edge computing, and not by a cloud-only architecture.

Chapter 8 examines security goals that every designer should aim to achieve. Next, an overview of the most important security challenges, threats, attacks, and vulnerabilities faced by IoT and wearable devices is provided. Finally, a list of security design consideration and best practices and ideas that have historically worked are discussed.

Chapter 9 first addresses the privacy issues and concerns arising from IoT and Wearable Technology, including those related to health data and data collected from children. The chapter next turns to safety and health issues then discusses the social and psychological impacts of these technologies. Finally, this chapter examines regulatory actions in the United States performed by the federal government, including the Federal Trade Commission (FTC), National Telecommunications & Information Administration (NTIA), as well as the ones performed by the private sector practicing self-regulation within the industry. As a means of comparison, this chapter next discusses the regulatory actions taken by the European Union.

Finally, the aim of Chapter 10 is to apply the knowledge learned in previous chapters to design two complete IoT and Wearable Technology products from scratch. This chapter will take the reader from concept and engineering requirements through breadboarding, microcontroller coding, PCB design, PCB printing, surface mount considerations all the way to a finished product.

Haider Raad, Ph.D.
Xavier University, USA

Acknowledgment

The author would like to thank Scott Tattersall and Mustafa Kamoona for their efforts in co-authoring Chapter 10 of this book. He also would like to thank Colin Terry for his help in developing the book's solution manual, and all the book's reviewers for their constructive feedback.

1

Introduction and Historical Background

1.1 Introduction

We live in a connected world where billions of computers, tablets, smartphones, buildings, wearable gadgets, medical devices, gaming consoles, and other smart items are constantly acquiring, processing, and delivering information. In the midst of this, the topics of Internet of Things (IoT) and wearable technology have begun to enjoy tremendous popularity thanks to the rapid advancements in digital systems, communication and information technologies, and innovative manufacturing and packaging techniques.

IoT and wearable devices have managed to swiftly gain a notable position in the consumer electronics market and are now making their way to become the new go-to technologies to address the needs of many industries. For instance, the retail industry has begun to use innovative inventory tracking and theft prevention devices based on IoT. The smart tag system enables self-checkout and allows business owners to track and manage their inventory in real time. The construction and mining sectors are increasingly investing in the use of wearable devices for hazard and health management by monitoring the environmental quality, detecing of approaching hazards, and assessing the physiological parameters of workers. IoT-based solutions are already being utilized in agriculture. Such systems are used to evaluate field variables such as soil condition, atmospheric parameters, and biological signals from plants and animals. They are also used to analyze and control variables such as temperature, PH levels, humidity, and vibrations, while being transported. Moreover, wearables are emerging as a solution to make healthcare accessible in remote areas (i.e. telemedicine). A plethora of wearable devices is already being used by medical professionals to aggregate physiological, behavioral, and biochemical data for diagnosing, treating, and managing chronic diseases.

Fundamentals of IoT and Wearable Technology Design, First Edition. Haider Raad.

Published 2021 by John Wiley & Sons, Inc.

IoT and wearable technology are all about enabling connectivity among humans and objects and unobtrusively delivering information and services to the right person at the right time. Their potential benefits are virtually limitless, and their applications are radically changing the way we live and are opening new opportunities for growth and innovation. This is just the tip of a massive iceberg.

This chapter presents a general overview and characteristics of IoT and wearable technology followed by a historical background; and finally, challenges that face these technologies are discussed.

1.1.1 IoT and Wearables Market Size

With around 18 billion devices connected to the Internet as of 2018, Cisco Systems predicts that this number will reach 50 billion by 2020. A recent report by the United Kingdom government speculates that this number could be even higher, in the range of 100 billion "things" connected.

A recent market analysis reports that the combined markets of IoT and wearables will grow to about $520 billion in 2021, compared to $235 billion spent in 2017. Another report indicated that the global shipments of wearables reached 49.6 million units in 2019, 55.2% up from 2018, with smart watches and wristbands continuing to dominate the wearables landscape, accounting for 63.2% of all devices shipped in that year. It is anticipated that the global wearables market share will exceed $51.50 billion by 2022.

What these numbers mean is that IoT and wearables will primarily shift the way people and businesses interact with their surroundings. Management and monitoring smart objects and systems using real-time connectivity will enable an entirely new level of data-driven decision making. This in turn will yield to optimized processes and deliver new services that save time and money for both people and enterprises.

1.1.2 The World of IoT and Wearables

The capability of IoT and wearable devices to communicate, process, and exchange information is the basis of their operation dynamics. These devices encompass a wide range of electronic components, sensors, actuators, computing technologies, in addition to communication and information protocols. Such technologies and protocols include but are not limited to wireless sensor networks, edge and cloud computing, big data analytics, embedded systems, security architectures, web services, and semantic search engines. However, this is also true for several other existing devices and technologies, so what makes them different?

1.1.2.1 What Is an IoT Device?

We have been connecting devices and "things" to the Internet and other networks for decades. Technologies such as automated teller machine (ATM), wireless sensor networks (WSN), machine to machine (M2M), and other connected devices are not new at all. However, this does not mean by definition that all these systems and devices are part of what we know today as the IoT. In other words, not all connected devices are IoT devices; however, all IoT devices are connected devices. Furthermore, in IoT we use the Internet Protocol (IP), IPv6,[1] in particular. Hence, we only pronounce the word Internet of Things when "things" are uniquely addressable.

There are many IoT definitions, and there isn't a universal one. It depends from which angle it is being looked at: Technology angle, application angle, or the industry angle.

However, from a general perspective IoT could be defined as the interconnection of devices with embedded sensing, actuating, and communication capabilities. Data in IoT are collected, processed, coordinated, and communicated through embedded electronics, firmware, and communication technologies, protocols, and platforms.

1.1.2.2 Characteristics of IoT Systems

We can also define the IoT as a network of connected devices, which have embedded and/or equipped with technologies that enable them to perceive, aggregate, and communicate meaningful information about the environment in which they are placed in, and/or themselves. The key characteristics of IoT from a broad-view perspective are as follows:

Unique Identity: As mentioned above, IoT is a network of connected devices with unique identifiers. It should be noted, however, that not all IoT devices are directly connected to the Internet. It is not always possible or even desirable to do so. In fact, a good number of IoT devices in a smart home or a factory setting communicate via a non-IP link such as ZigBee or low-power wide-area network (LoRaWAN), which enables these devices to communicate over distances with gateways that interface to standard IP networks. It should also be noted that when such devices use IP, it does not by default mean they are using the public Internet. There could be home and enterprise networks that use IP with data traffic that may never touch the public Internet. Also, even if there is a cluster

1 **IPv6** (Internet Protocol version 6) is a network protocol launched in 2012 that enables data communications over a packet-switched network. The explosive growth in connected devices has driven a need for additional unique IP addresses. The previous standard IPv4 supported a maximum of approximately 4.3 billion unique IP addresses, while IPv6 supports a theoretical maximum of 3.4×10^{38} addresses.

of non-IP devices communicating with an aggregation gateway, beyond that gateway, the expectation is that the traffic will be IP-based. Hence, all the "nodes" of the IoT are expected to create some sort of IP traffic, whether directly, or through some gateway.

Sensing and Actuating: Sensors and actuators are two crucial elements in IoT systems. Sensors are used to perceive and gather information about some dynamic activity (pressure, temperature, altitude, etc.). The collected information is resulted from the interaction of the sensor with the environment.

A more general expression for a sensor is a transducer. A transducer is any device that converts one form of energy into another. A microphone, for instance, is a transducer that takes sound energy and converts it to electrical energy in a useful manner for other components in the system to correlate with.

An actuator is another type of transducer that is found in the majority of IoT systems. Actuators operate in the reverse manner as sensors. They typically take a form of energy and convert it into a physical action. For example, a speaker takes an electrical signal and converts it into a diaphragm vibration which replicates an audio signal.

Connectivity, Communication, and Data Distribution: IoT devices are connected to the Internet either directly or through another device (gateway) where network connections are used for transporting data and interacting with users. Also, these devices allow users to access information and/or control devices remotely using a variety of communication protocols and technologies.

Automation: Regardless of the application, most IoT devices are about automation, such as in industrial automation, business process automation, or home automation. Thus, such devices can generate, exchange, and produce data with minimal or no human intervention.

Intelligence: Intelligence in IoT lies in the knowledge extraction from the generated data and the smart utilization of this knowledge to solve a challenge, automate a process, or improve a situation. There is no real IoT benefit without artificial intelligence, machine learning, Big Data[2] analytics, and cognitive algorithms.

Figure 1.1 depicts the abovementioned IoT characteristics.

1.1.2.3 What Exactly Is a Wearable Device?

The term "wearable devices" generally refers to electronic and computing technologies that are incorporated into accessories or garments which can comfortably be

---

2 **Big Data** is a term that describes the diverse and large volumes of data that grow at an increasing rate. It encompasses the volume of information, the speed at which it is created, aggregated, and collected, and the variety of points being covered. Typically, Big Data comes from multiple sources and takes multiple formats.

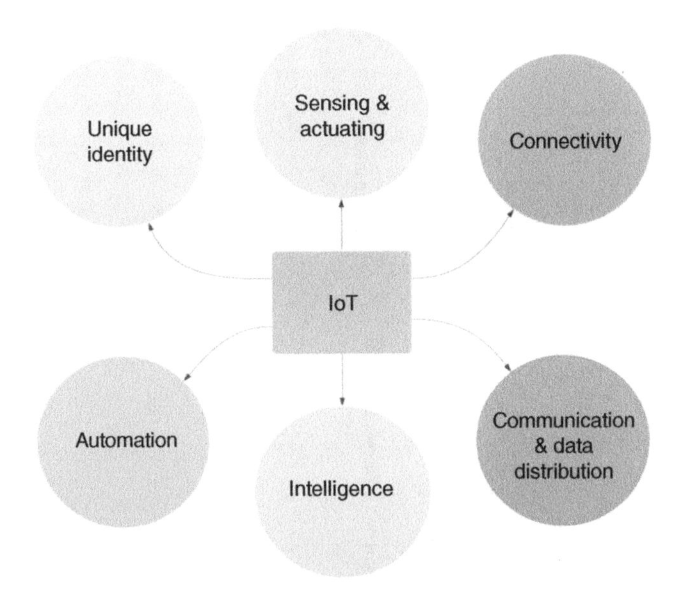

Figure 1.1 Characteristics of the Internet of Things.

worn on the user's body. These devices are capable of performing several of the tasks and functions as smartphones, laptops, and tablets. However, in some cases, wearable devices can perform tasks more conveniently and more efficiently than portable and hand-held devices. They also tend to be more sophisticated in terms of sensory feedback and actuating capabilities as compared to hand-held and portable technologies. The ultimate purpose of wearable technology is to deliver reliable, consistent, convenient, seamless, and hands-free digital services.

Typically, wearable devices provide feedback communications of some sort to allow the user to view/access information in real time. A friendly user interface is also an essential feature of such devices, so is an ergonomic design. Examples of wearable devices include smart watches, bracelets, eyewear (i.e.: glasses, contact lenses), headgears (i.e.: helmets), and smart clothing. Figure 1.2 depicts the most important possible forms of wearable devices.

While typical wearable devices tend to refer to items which can be placed external to the body surface or clothing, there are more invasive forms as in the case of implantable electronics and sensors. In the author's opinion, invasive implantables, i.e. ingestible sensors, under the skin microchips, and smart tattoos, which are generally used for medical and tracking purposes, should not be categorized as wearables since they have different mechanisms and operation requirements. The reader should seek other resources which are dedicated to the design and prototyping of such devices.

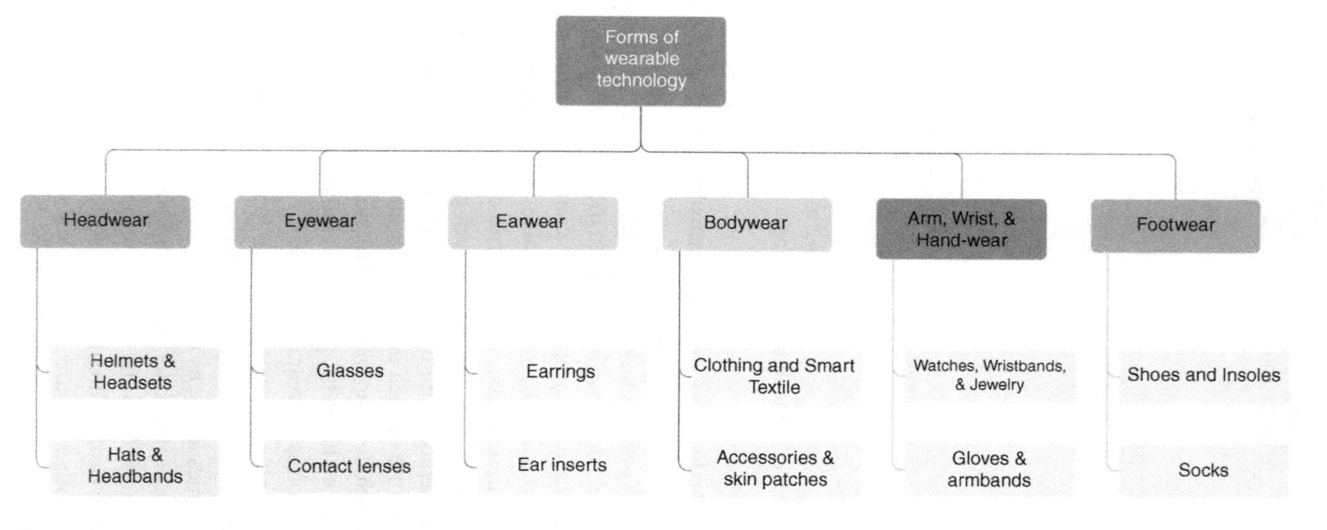

Figure 1.2 Forms of wearable technology.

1.1.2.4 Characteristics of Wearable Devices

The uses of wearables are far reaching and have exciting potentials in the fields of medicine, well-being, sports, aging, disabilities, education, transportation, enterprise, and entertainment. The main objective of wearable technology in each of these fields is to smoothly incorporate functional and portable electronics into the users' daily routines. Prior to their existence in the consumer market, wearables were primarily employed in the fields of military technology and health sector.

Generally speaking, wearables share many aspects of the sensing, connectivity, automation, and intelligence characteristics with IoT devices. However, there are a few major differences worth highlighting which will be discussed in the following sections.

Form factor is a hardware design aspect in electronics packaging which specifies the physical dimensions, shape, weight, and other components specifications of the printed circuit board (PCB) or the device itself. Although wearable devices have a small form factor in general, it is practically dependent on the type and the way they are worn (rings and wristbands, as opposed to glasses and clothing).

Smaller form factors may offer reduced usage of material, easy handling, and simpler logistics; however, they typically give rise to higher design and manufacturing costs in addition to signal integrity issues and maintenance constraints.

Moreover, durability, comfort, aesthetics, and ergonomic factors are important when it comes to designing a wearable device. Weight, shape, color, and texture must be carefully considered. The general characteristics of wearable technology are presented in Figure 1.3.

1.1.2.5 IoT vs. M2M

M2M describes the technology that enables the communication between two or more machines. With M2M, one could connect machines, devices, and appliances in a wired or wireless fashion via a variety of communications techniques to deliver services with limited human intervention.

The difference between machine to machine (M2M) and IoT can be confusing to many. In fact, the misconception that M2M and IoT are the same has been a continuing subject of debate in the realm of tech industry.

Both M2M and IoT are connectivity solutions that provide remote access to machine data. They both have the capability of exchanging information among machines without human intervention. Thus, the two terms have been mistakenly interchanged often. However, M2M is a predecessor to IoT and had revolutionized enterprise operations by enabling them to monitor and manage their machines and hardware components remotely. M2M set the underlying basis of machine connectivity on which IoT built upon. Nevertheless, IoT is the ultimate manifestation when it comes to connectivity.

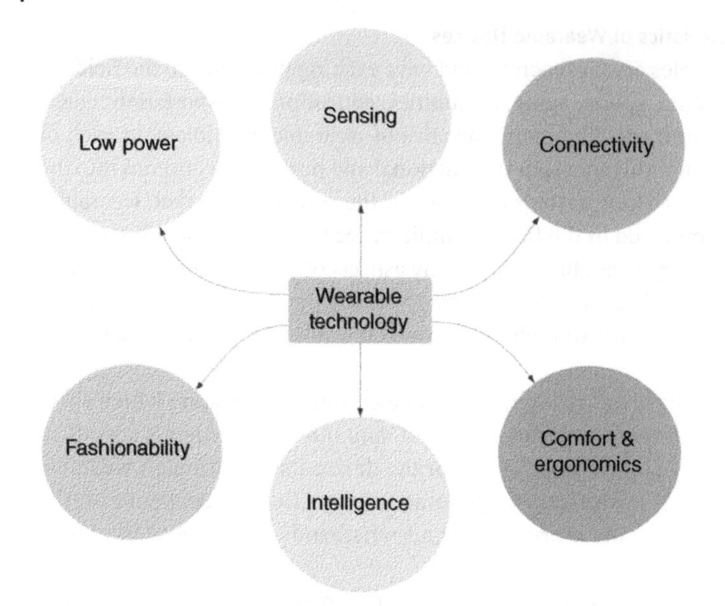

Figure 1.3 Characteristics of wearable technology.

The main objective of M2M is to connect a machine/device to another machine (typically in an industrial setting) via cellular or wired network so that its status can be monitored and its data can be collected, remotely. IoT is more of a universal market technology that aims at serving consumers, industries, and enterprises. Consumer IoT connects users to their devices and enables remote access. On the other hand, enterprise and industrial IoT take it further by allowing tracking, control, and management.

IoT and M2M diverge immensely when it comes to the way they access devices remotely. M2M relies on point-to-point communications enabled by dedicated hardware components integrated within the machine. The communication among these connected machines is made possible via wired or conventional cellular network and dedicated software. IoT, on the other hand, typically uses IP networks and integrates web applications to interface device/machine data to a middleware, and in the majority of cases, to cloud.

It is worth noting that IoT is intrinsically more scalable than M2M since cloud-based architectures do not need additional hard-wired connections and subscriber identification modules (SIM) which are required in M2M.

1.1.2.6 IoT vs. Wearables

Despite the commonalities, it is clear that there are substantial differences when we speak about wearable technology in the context of fitness trackers as opposed to when IoT is used in the context of manufacturing processes or smart cities. In fact, many experts in the field argue that wearables fall under the umbrella of IoT. One key difference worth highlighting here is that most wearables rely on a gateway device, such as a smartphone, for configuration and connectivity, and in most cases to enable features and process data. It is this M2M aspect that makes wearables a separate class of devices, and that's why we prefer to treat these as two technologies with two sets of characteristics.

It is also worth noting that not all wearable devices require connectivity, for example, a simple pedometer and an ultraviolet monitor could operate offline. Other wearables require minimal connectivity only.

Although IoT and wearable devices have a lot in common in terms of design aspects, components, and technologies and protocols used, there are still some real differences that architects and designers need to be aware of. Figure 1.4 shows a table summarizing the main differences between M2M, IoT, and Wearable Technology.

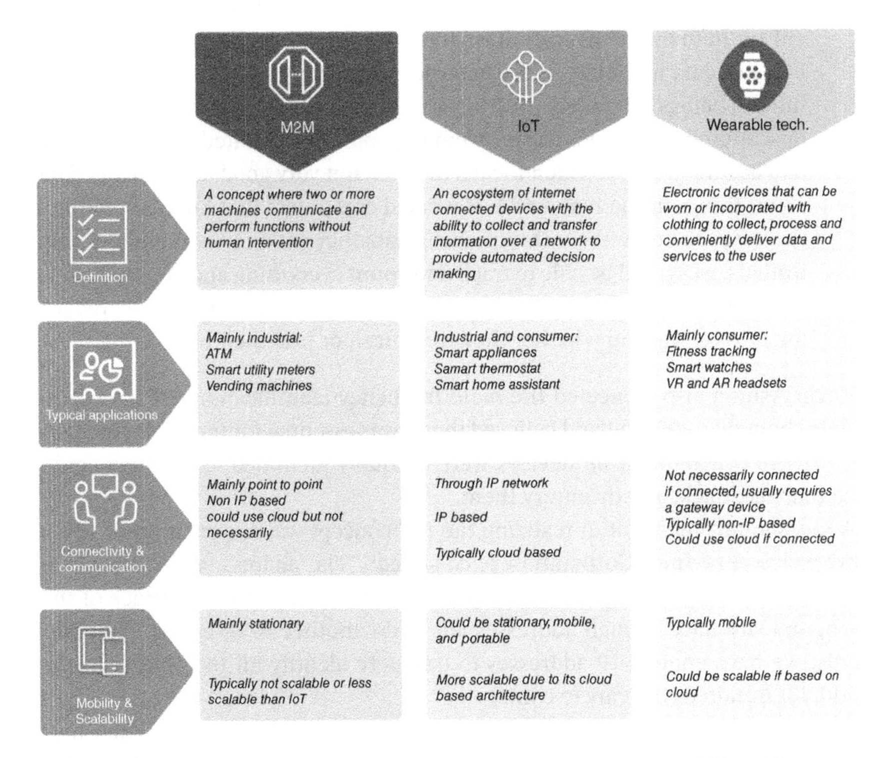

Figure 1.4 A summary of the main differences between M2M, IoT, and Wearable Technology.

1.1.3 IoT: Historical Background

The term "IoT" has not been around for so long. However, the idea of machines communicating with one another has been brewing since the telegraph was developed in early 1800s.

The first wireless transmission over a radio took place in 1900, bringing about endless innovations. This crucial ingredient of the future IoT was complemented by the inception of computers in the 1950s.

An essential component of the IoT is the Internet itself which was initiated as part of the Defense Advanced Research Projects Agency (DARPA) in 1962 and then progressed into ARPANET in 1969. In the 1980s service providers started promoting the commercial use of ARPANET, which matured into today's Internet.

The term IoT was not officially coined until 1999 when Kevin Ashton, the executive director of Auto-ID Labs at MIT, was the first to describe the Internet of Things in a presentation for Procter & Gamble. During his speech, Ashton stated:

> Today computers, and, therefore, the Internet, are almost wholly dependent on human beings for information. Nearly all of the roughly 50 petabytes of data available on the Internet were first captured and created by human beings by typing, pressing a record button, taking a digital picture or scanning a bar code. The problem is, people have limited time, attention, and accuracy. All of which means they are not very good at capturing data about things in the real world. If we had computers that knew everything there was to know about things, using data they gathered without any help from us, we would be able to track and count everything and greatly reduce waste, loss and cost. We would know when things needed replacing, repairing or recalling and whether they were fresh or past their best.

Kevin Ashton also pioneered the radio-frequency identification (RFID) use in supply chain management and believed that it was essential for the deployment of the IoT. He concluded if all devices were uniquely identified, computers could then manage, track, and inventory them.

A foundational element in realizing the IoT concept was the creation of IPV6. Steve Leibson of Intel Corporation once stated: "The address space expansion means that we could assign an IPV6 address to every atom on the surface of the earth, and still have enough addresses left to do another 100+ earths." In other words, we have enough IP addresses to uniquely identify all the objects in the world, for hundreds of years to come.

One of the early examples of an Internet of Things is from 1982, when four students from the School of Computer Science department installed switches in a Coca Cola machine at the Carnegie Melon University. The students would connect by ARPANET to the appliance and remotely check the availability of the drink, and if it was cold, before making the trip to the machine. This experiment had inspired numerous inventors around the world to devise their own connected appliances.

After the invention of the World Wide Web by the British scientist Tim Berners-Lee in 1989 and the launching of commercial Global Positioning System, inventors had been able to develop interconnected devices way more efficiently. One of the first examples was an Internet-connected toaster introduced by John Romkey in 1990, which is considered by many as the first "real" IoT device.

In 1991, two academicians who worked at the computer laboratory in the University of Cambridge set up a camera to provide live picture of a coffee pot (known as the Trojan Room coffee pot) to all desktop computers on the office network to save people working in the building time and from getting disappointed of finding the coffee pot empty after making the trip. This invention was a true inspiration for the world's first webcam. A few years later, the coffee pot was connected to the Internet and gained international fame until it was retired in 2001.

In the year 2000, LG announces Internet Digital DIOS, the world's first Internet-enabled refrigerator. The refrigerator became a buzzword despite its commercial failure.

In 2004, Walmart Inc. required its top suppliers to assign RFID tags to cases and pallets in place of barcodes by 2005 to enhance their supply chain operations. The suppliers were unhappy with the new requirements as Electronic Product Code (EPC) tags were pricey and seemed unnecessary. Walmart, subsequently, offered the suppliers to disclose point of sales information which led to a decreased merchandise thefts and labor requirements. Currently, EPC is one of the international standards, connecting billions of "things" worldwide.

The year 2005 witnessed the first Internet-connected robot, the Nabaztag rabbit. The bunny-shaped robot is capable of gathering weather reports, news, and stock market updates through Wi-Fi connectivity and reading them to the consumer. Despite its retirement in 2015 due to technological impediments, Nabaztag proved that IoT can be integrated into everyday lives.

The First International Conference on the Internet of Things took place in Zurich, Switzerland, in 2008. The event was the first conference of its kind with participants from 23 countries. The same year marked the first time where more "things" are connected to the Internet than people. A year later, Google started the first testing of self-driving cars while St. Jude Medical officially became an adopter of IoT for healthcare.

The year 2010 marks the first time IoT was recognized on a governmental level where China's head of government Wen Jiabao decided to pay special attention to IoT as one of the remedies to his country's financial crisis and adopting it across top strategic industries. The same year also marks the first implementation of machine learning techniques in IoT devices. Nest smart thermostat was the first IoT product to adapt to the user's habit and thus optimizing the air conditioning schedule.

By the year 2013, IoT had evolved into a system that utilizes multiple technologies, ranging from embedded systems and wireless communication to electromechanical sensors and control systems.

In 2014, Google Inc. acquires Nest after spotting the potential behind IoT and smart home devices in particular. Moreover, Google's self-driving car prototype was ready for testing on public roads but would not perform the official test drive until the following year.

On the 6th November of the same year, Amazon releases Echo, the first commercially successful voice-controlled ambient device and IoT hub. It is also anticipated that Amazon's device will be one of the most disruptive technologies in the next generation of enterprise IoT solutions.

The Global Standards Initiative on IoT takes place in 2015. The event's main objective was to establish a unified approach to the development of IoT technical standards and to support the adoption of the technology, globally.

In 2016, the automotive giant General Motors invests $500 million in Lyft aiming at developing a network of self-driving cars. In the same year, Apple showcases HomeKit products at the Consumer Electronics show. HomeKit is a platform that allows developers to utilize a comprehensive list of software tools for smart home application. In the meantime Google releases Google Home, another smart ambient device competing with Amazon's Echo. This year also witnessed the emergence of the first IoT malware.

In 2017, Microsoft launches Azure IoT edge that allows IoT devices to deploy complex processing and analytics locally, while Amazon offers advanced security features, Google releases Cloud IoT Core which allows an easier connectivity to the cloud. Witnessing such initiatives from giant technology leaders, one can realize that IoT is here to stay.

1.1.4 Wearable Technology: Historical Background

The beginning of this decade has surely witnessed the increasing number of wearable devices where one can spot numerous variations of smart watches, health assistive gadgets, fitness trackers, and smart clothes on the shelves. The growing number of these sleek devices since then along with their expanding applications clearly indicates that wearables are thriving. But one may ask: When and how did it all begin?

Here, we are not discussing the first ubiquitous wearable technology: the eyeglasses, which dates back to the thirteenth century, nor the abacus ring which dates back to the early days of China's Qing dynasty in the seventeenth century. We are specifically addressing smart wearables that have digital computational power!

One may be surprised to learn that much of the history of wearables is found in a "smart" shoe used to cheat at roulette tables in casinos! In 1961, Edward Thorp and Claude Shannon[3] built computing devices that could predict where the ball would land on a roulette wheel which could improve the chances of winning a bet by up to 44%. Obviously, Thorp and Shannon's apparatus was not illegal at the time of invention. One of devices was concealed in a shoe, while the other in a pack of cigarettes. It is worth mentioning that Edward Thorp credits himself as the inventor of the first wearable computing device. Other variations of such apparatus were designed and built in the 1960s and 1970s targeting the casino business, perhaps the most widely known is "George," a shoe-based wearable device designed by Keith Taft who used his toes to operate it. The smart shoe was used to gain an advantage at Blackjack tables.

Previously, in 1938, Aurex Corp., a Chicago-based electronics firm, developed the first electronic hearing aid device, marking one of the first innovations in the biomedical wearables industry. In 1958, the world's first pacemaker was invented by Earl Bakken. One might argue that these are not "smart technologies" since they are not based on a digital computing system; however, they gave rise to their smart counterparts we know today.

On the other hand, the first "smart watch" was first launched in 1975 holding the brand name "Pulsar." The smart watch was primarily a wearable calculator that also tells time in a digital format. The Pulsar became a widely adopted gadget by electronics enthusiasts and math geeks all over the world! Despite their drastic popularity decline, these watches are still being produced by many manufacturers to this day (Figure 1.5).

Some might argue that the iconic Walkman music player was the first ever wearable technology that truly went mainstream. The Japanese brand SONY launched the Walkman in 1979 and was followed by a triumphant commercial success as it significantly transformed the music listening routines for millions of consumers around the world. SONY's Walkman production line was discontinued indefinitely in 2010 with over 220 million machines sold worldwide.

3 Claude Shannon is also known as the father of information theory with his legendary paper "A Mathematical Theory of Communication" published in 1948. He is also well known for founding the digital circuit design and cryptanalysis theories in 1930s when he was in his early twenties as a master's student at the Massachusetts Institute of Technology (MIT).
Edward Oakley "Ed" Thorp was an American mathematics professor at the University of California, Irvine between 1965 and 1977, author of the books (Beat the Dealer and Beat the Market), and blackjack player. He is best known as the "father of the wearable computer."

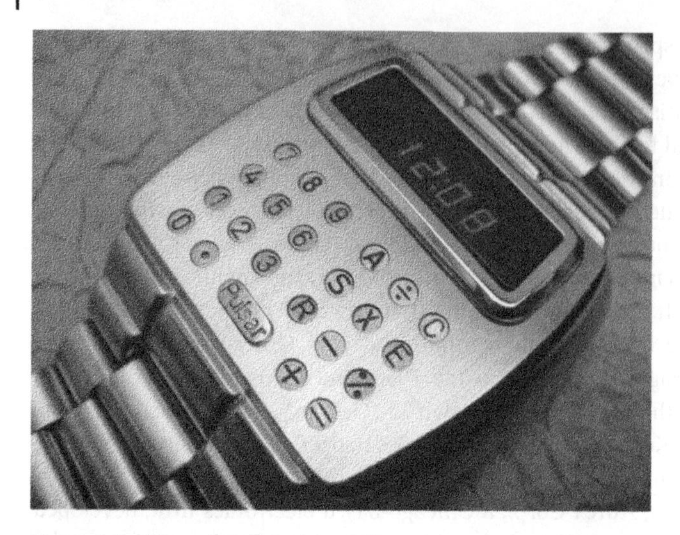

Figure 1.5 The pulsar calculator LED watch released in 1975. *Source:* Photo courtesy of Piotr Samulik.

In 1981, Steve Mann, a high school student, incorporated an Apple II (6502) computer into a steel-framed backpack to control a photography apparatus attached to a helmet. It is also worth mentioning that Steve Mann is also known for creating the first wearable wireless webcam in 1994 and as the first lifelogger.[4] He has also pioneered many innovations in the fields of wearable technology and digital photography (Figure 1.6).

In the realm of health care, the first practical and fully digital hearing aid device was invented by Engebretson, Morley, and Popelka. Their patent, "Hearing aids, signal supplying apparatus, systems for compensating hearing deficiencies, and methods" filed in 1984 served as the basis of all subsequent digital hearing aid devices, including those produced today.

The mid-1990s marked the brainstorming period for wearable technology where conferences and expos on wearables and smart textiles began to see a rise in popularity. The DARPA held its forward-thinking workshop in 1996 entitled "Wearables in 2005." One of DARPA's galvanizing predictions included computerized gloves that could read RFID tags. However, wearables were overshadowed by the smartphone revolution between the late 1990s and mid-2000s, smartphones were simply the consumer's gadget of choice, due to obvious reasons.

---

4 A **lifelogger** is a person who uses/wears a recording apparatus or a computer in order to capture a substantial portion of his/her life.

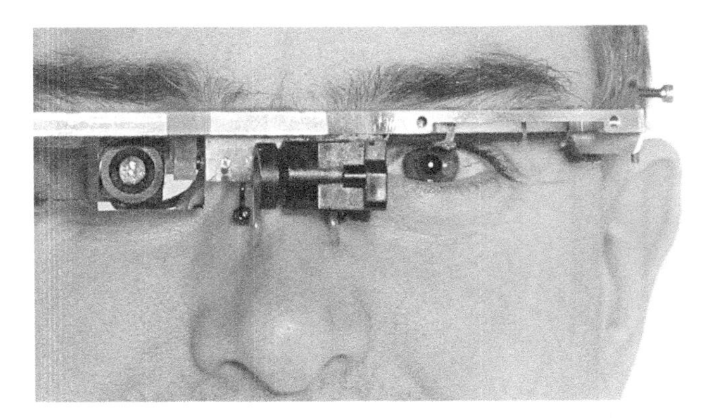

Figure 1.6 Steve Mann wearing one of his wireless wearable webcam. *Source:* Glogger, https://commons.wikimedia.org/wiki/File:SteveMann_with_Generation-4_Glass_1999.jpg. Licensed Under CC BY-SA 3.0.

In 2003, the Garmin Forerunner, a watch that tracks the user's performance, emerged which was immediately followed by popular fitness trackers we all know today such as the Nike+, Jawbone, and Fitbit.

Toward the end of 2000s, several Chinese companies started producing Global System for Mobile (GSM) phones integrated within wristbands and equipped with mini displays. On the other hand, the first smart watch, Pebble, came to the scene in 2012, followed by the much-hyped Apple Watch in 2014.

Future wearables may enable new functions and services that one could barely imagine, but it is clear to see how early wearables evolved into the fascinating devices we enjoy today.

1.1.4.1 The Wearables We Know Today

One of the most publicized wearables today is the Apple Watch. The watch incorporates activity and health tracking capabilities with other Apple applications. The primary goal of the Apple Watch was to improve the way users interact with their iPhones and introduce extra convenience (Figure 1.7). The birth of the Watch began when Kevin Lynch was recruited by Apple to create a wearable technology for the wrist. He said: "People are carrying their phones with them and looking at the screen so much. People want that level of engagement. But how do we provide it in a way that's a little more human, a little more in the moment when you're with somebody?".

Resonating with today's technological advancement, modern consumers take an active role in utilizing wearables to track and record data of their active lifestyles. Nowadays, wearable fitness and health trackers are capable of monitoring

Figure 1.7 The Apple watch. *Source:* Photo courtesy of Apple Inc.

the user's biometric data including heart rate, blood pressure, temperature, calories, and sleep patterns.

Another hot wearable, Fitbit, is capable of measuring personal fitness metrics such as the number of steps walked or climbed, heart rate, sleep patterns, and even stress levels (Figure 1.8).

On the other hand, many argue that the most innovative wearable device of the decade is the Google Glass, which is fundamentally a pair of glasses equipped with a built-in microprocessor and a bundle of peripherals such as a mini display embodies by a 640×360 pixels prism projector that beams out a viewing screen into the user's right eye, a gesture control pad, a camera, and a microphone. The Glass runs a specially designed operating system (Glass OS) and has 2 GB of RAM and 16 GB of flash storage, in addition to a gyroscope, an accelerometer, and a light sensor. Through such peripherals, the user could connect to his/her smartphone, access mobile Internet browser, camera, maps, and other apps by voice commands. It accesses the phone through Wi-Fi and Bluetooth which are enabled by the wireless service of the user's mobile phone.

Figure 1.8 Fitbit Surge smart watch fitness tracker. *Source:* Photo courtesy of Fitbit©.

Google released the consumer version of Glass in 2013 amid much fanfare, but it failed to gain commercial success. The Glass also faced serious criticism due to concerns that its use could violate current privacy laws. In 2017, Google launched the Glass Enterprise Edition after deciding that the Glass was better suited to workers who need hands-free access to information, such as in health care, manufacturing, and logistics. In 2019, Google has announced a new version of its Enterprise Edition which has an improved processor, camera, charging unit, and various other updates.

One can imagine a considerable number of applications this technology is capable of creating. In fact, the Glass is already being utilized in a number of areas once considered "futuristic." For example, Augmedix, a San Francisco based company, developed a Glass app that allows physicians to livestream the patient visit. The company claims that electronic health record problems will be eliminated, and their system would possibly save doctors up to 15 hours a week.

In 2013, Rafael Grossmann was the first surgeon to demonstrate the use of Google Glass during a live surgical procedure. In the same year, the Glass was used by an Ohio State University surgeon to consult with another colleague, remotely.

Obviously, such technology could have a positive impact on the lives of people with disabilities. For example, one application is designed to enable parents to swiftly access sign language dictionary through voice commands in order to communicate effectively with their deaf children.

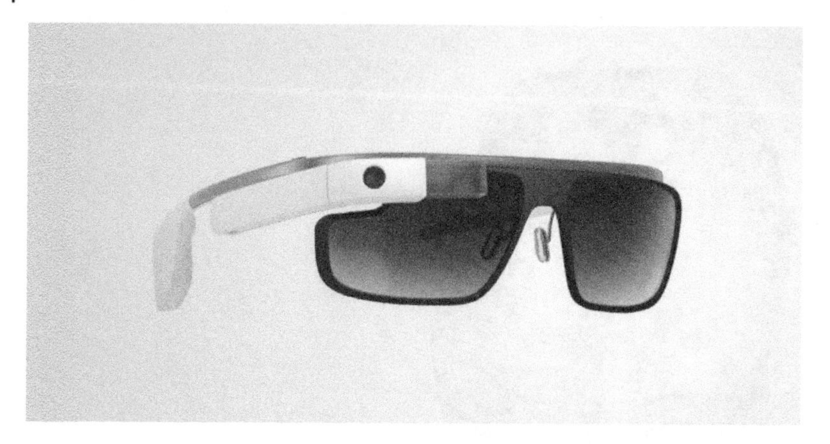

Figure 1.9 Explorer edition of Google Glass©. *Source:* Photo courtesy of Google Inc.

Using a smart glass technology in the tourism and leisure industry, the experience of tourists could be substantially improved. Attractions and museum tours can be immensely enhanced by displaying text or providing audible information when recognizable buildings, sculptures, and artwork are detected. Users will also be able to capture photographs and videos more conveniently, i.e. via voice command or a wink of an eye. Another helpful application dedicated to break the language barriers when traveling provides instantaneous translation. Any text visible to the Glass field of view can be translated via voice commands (Figure 1.9).

Boeing is using the Glass to help their assembly crew in the connecting aircraft wire harnesses, which is a very lengthy process that requires a high volume of paperwork. The crew now could have a hands-free access to the needed information using voice commands.

Stanford University is conducting a breakthrough research dedicated to help autism patients read the emotions of others using the Glass by utilizing facial recognition software to determine the emotions expressed on the people's faces projected within the display.

In 2014, Novartis and Google X (now X)[5] started the testing of a smart contact lens in the field of telehealth.[6] The lens is equipped with a miniaturized glucose sensor that continuously tracks blood sugar levels through the diabetic patient's tears and communicates the data to a smartphone through a wireless module. In 2018, Verily

---

5 X is an American semi-secret R&D center founded by Google in 2010 with the name Google X. The company's first project was Google's self-driving car. It is located about 0.5 mile from Google's headquarters in Mountain View, California.

6 Telehealth, also known as Telemedicine, is providing healthcare services from a distance through the use of telecommunication and information technologies. It came as a solution to improve access to healthcare services that would not be readily available, especially in remote regions.

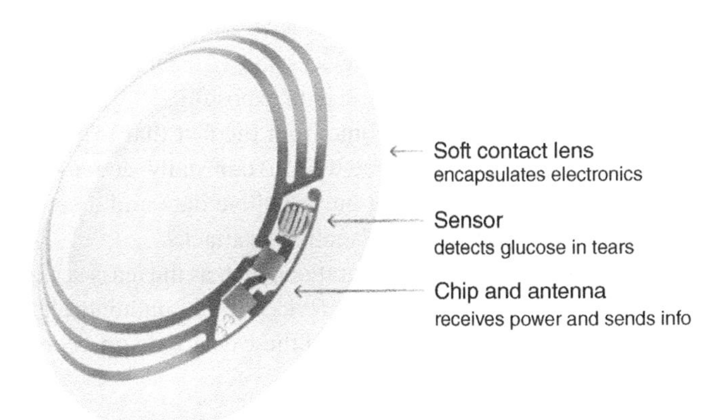

Figure 1.10 Infographic photo of the Google Smart Lens©. *Source:* Photo courtesy of Google Inc.

(a former division of Google X) announced that the lens project has been dismissed due to the lack of correlation between blood glucose and tears (Figure 1.10). However, competitors started to take advantage of Google's lens failures to work on developing their own smart eye wearables. For example, EPGLMed is working with Apple, to develop a smart lens that corrects vision on-demand by changing the curvature of the lenses through a smartphone app.

In summary, the applications of wearable technology are extremely powerful and they are evolving rapidly. It is crystal clear that this technology is here to stay.

1.1.5 Challenges

While the IoT and wearable technology continue to transform our lives in the twenty-first century, significant challenges that could stand in the way of realizing its full potential are coming to light. Below are the major challenges that require full attention:

1.1.5.1 Security

Security is one of the cornerstones of the Internet and is the most significant challenge for IoT and wearable devices. The hacking of fitness trackers, security cameras, baby monitors, and other abuses has drawn the attention and serious concerns of major tech firms and government agencies across the world.

While security considerations are not new in the realm of information technology, the characteristics of many IoT and wearable technology deployments introduce new and unprecedented security challenges. Addressing these challenges and ensuring secure IoT and wearable products and services must be a top priority. As these technologies are becoming more pervasive and integrated into our daily lives, users need to be assured that these devices and associated data are secure from vulnerabilities such as cyber-attacks and data exposure.

The more consequential shift in security will come from the fact that IoT and wearable technology will become more integrated into our daily activities. Concerns will no longer be limited to protecting our sensitive data and assets. Our own lives and health can become the target of malicious attacks.

This challenge is further amplified by other considerations such as the mass-scale production of identical devices, the ability of some devices to be automatically paired with other devices, and the potential deployment of these devices in unsecure environments.

1.1.5.2 Privacy

While many of the emerging IoT and wearable technologies are giving rise to a spectrum of new applications and innovative uses, as well as promising and attractive benefits, they also pose privacy concerns that are largely unexplored. In fact, a new research area concerning the security and privacy of these technologies has recently emerged. Additionally, the need for the majority of wearable devices and a good number of IoT systems to interact and share data with an access point (i.e. a smart watch to smartphone, medical monitoring device to a home server, smart bulb to an ambient home assistant) along with other sensors and peripherals would certainly create a new class of privacy and security hazards.

Some IoT and wearable devices deploy various sensors to collect a wide spectrum of biological, environmental, behavioral, and social information from and for their users. Clearly, the more these devices are incorporated into our daily lives, the greater the amount of sensitive information will be transported, stored, and processed by these devices, which also elevate privacy concerns.

Moreover, integrated voice recognition or monitoring features are continuously listening to conversations or video record activities and selectively transmit such potentially sensitive data to a cloud service for processing, which sometimes involves a third party. Handling and interacting with such information unveil legal and regulatory challenges facing data protection and privacy laws.

One specific privacy concern associated with the emerging smart glasses is that they allow users to simultaneously record and share images and videos of people and their activities in their range of vision, in real time. This problem will soon be intensified when such devices are integrated with facial recognition programs which will allow users to see the person's name in the field of view, personal information, and even visit their social media accounts.

1.1.5.3 Standards and Regulations

The lack of standards and best practices documentations poses a major limitation to the potential of IoT and wearable devices. Without standards to guide manufacturers and developers, these products that often operate in a disruptive manner would lead to interoperability issues and might have negative impacts if poorly designed and configured. Such devices can have adverse consequences on the network resources they connect to and the broader Internet. Unfortunately, most of this comes down to cost constraints and the pressuring need to release a product to the market quicker than competitors. Moreover, there is a wide range of regulatory and legal questions surrounding the IoT and wearable technology, which require thoughtful consideration.

Legal issues with IoT and wearable devices may include conflicts between governmental surveillance and civil rights; policies of data retention and destruction; legal liability/penalty for unintended uses; and security breaches or privacy abuses. Furthermore, technology is advancing much faster than the associated policy and regulatory environments which might render policies and regulations to be inappropriate.

Big data presents another serious challenge. The analysis, extraction, manipulation, storage, and processing of substantial amounts of data may pose other legal problems as in profiling, behavior analysis, and monitoring. Big data may require new protection policies, international coordination, and infrastructure management, among others.

Furthermore, the cloud and even the Internet itself are not tied to one specific geographic location. Moreover, the sheer amount of IoT and wearable devices originate from a number of different sources, including international partners and vendors, which makes it impossible for a localized regulatory authority to enforce quality control or standardized tests.

As yet, these challenges have been minimally acted upon by policy makers. However, they reflect a pressing necessity to seek government solutions to both pronounce the strengths of these technologies and deploy policies to minimize their risks.

1.1.5.4 Energy and Power Issues

The increase in data rates and the number of Internet-enabled services and the exponential growth of IoT and wearable devices are leading to a substantial increase in network energy consumption.

Moreover, the push toward smaller size and lower power is creating more signal and power integrity problems in IoT and wearable devices. Common issues include mutual coupling, distortion, excessive losses, impedance mismatch, and generator noise. Failure to deal with these issues can have detrimental effects on these devices.

1.1.5.5 Connectivity

According to recent research reports, around 22 billion IoT and wearable devices will be connected to the Internet by 2020. Thus, it is just a matter of time before users begin to experience substantial bottlenecks in IoT connectivity, proficiency, and overall performance.

Currently, a big percentage of connected devices rely on centralized and server/client platforms to authenticate, authorize, and connect additional nodes in a given network. This model is sufficient for now, but as additional billions of devices join the network, such platforms will turn into a bottleneck. Such systems will require improved cloud servers that can handle such large amounts of information traffic. This is already being addressed by the academic and industrial community which is pushing toward decentralized networks. With such networks, some of the tasks are moved to the edge, such as using fog computing, which takes charge of time-sensitive operations (this will be discussed in detail in chapter 7), whereas cloud servers take on data assembly and analytical responsibilities.

1.2 Conclusion

IoT and wearable devices are enabled by the latest developments in smart sensors, embedded systems, and communication technologies and protocols. The fundamental premise is to have sensors and actuators work autonomously without human involvement to deliver a new class of applications. The recent technological revolution gave rise to the first phase of the IoT and wearable devices, and in the next few years, these devices are expected to bridge diverse technologies to enable novel applications by connecting physical objects together in favor of intelligent decision making.

Benefits are substantial, but so are the challenges. This will require businesses, governments, standards bodies, and academia to work together toward a common goal.

In short, IoT and wearable technology are representative icons of the most recent industrial revolution. Given that we advance and evolve by transforming data into information, knowledge, then into wisdom, these technologies have the potential to change the world as we know it today, in new and exciting ways.

Problems

1 What are the main differences between IoT and wearable technology?

2 What is it meant by "things" in Internet of Things?

3 What are the main differences between IoT and M2M?

4 Can you think of other potential challenges found in IoT and wearable technology other than the ones mentioned in this chapter?

5 Give examples of wearable devices/applications that do not require Internet connectivity.

6 List five real-world examples of smart clothing.

7 List five real-world examples of the headwear form in wearable technology.

8 List four components common between IoT and wearable devices (an application of your choice).

9 Are wearable devices a form of M2M? Why?

10 If you are asked to add more somewhat essential characteristics to IoT, what would they be? Why?

Interview Questions

1 In simple words, explain the term IoT.

2 Who are the key players in the field of IoT?

3 Who are the key players in the field of wearable technology?

4 What is M2M? Where does IoT intersect with M2M?

5 How is wearable technology expected to have an impact on our daily life?

6 How is 5G technology going to affect the deployment of IoT?

7 What will happen in terms of jobs losses and required skills as IoT makes devices more intelligent?

8 How would wearable technology affect businesses?

9 What is the difference between the "Things" in "Internet of Things" and sensors?

10 What is the connection of IoT to Big Data?

Further Reading

Aazam, M. and Huh, E.-N. (2014). Fog computing and smart gateway based communication for cloud of things. *Proceedings of the 2nd IEEE International Conference on Future Internet of Things and Cloud (FiCloud '14)*, Barcelona, Spain (August 2014), pp. 464–470.

Atzori, L., Iera, A., and Morabito, G. (2011). SIoT: giving a social structure to the Internet of Things. *IEEE Communications Letters* 15 (11): 1193–1195.

Bertolucci, J. (2010). Reliability report card: grading tech's biggest brands. *PC World* 27 (2): 82–92. Chan, J. November 4.

Erfinder, A., Engebretson, A.M., Morley, R.E. Jr., and Popelka, G.R. (1984). Hearing aids, signal supplying apparatus, systems for compensating hearing deficiencies, and methods. US Patent 4548082.

Guo, B., Zhang, D., Wang, Z. et al. (2013). Opportunistic IoT: exploring the harmonious interaction between human and the internet of things. *Journal of Network and Computer Applications* 36 (6): 1531–1539.

Hayes, A. (2017). A brief history of wearable computing. Bradley Rhodes - MIT Media Lab, MIT Wearable Computing Project. https://www.media.mit.edu/wearables/lizzy/timeline.html (accessed January 2017).

Holland, J. (2016). *Wearable Technology and Mobile Innovations for Next-Generation Education*. Hershey, PA: IGI Global, ISBN-13:9781522500698.

Khaleel, H.R. (2014). *Innovation in Wearable and Flexible Antennas*. Southampton, UK: WIT Press.

Liang, G., Cao, J., and Zhu, W. (2013). CircleSense: a pervasive computing system for recognizing social activities. *Proceedings of the 11th IEEE International Conference on Pervasive Computing and Communications (PerCom '13)* (March 2013). San Diego, CA: IEEE, pp. 201–206.

Mashal, I., Alsaryrah, O., Chung, T.-Y. et al. (2015). Choices for interaction with things on Internet and underlying issues. *Ad Hoc Networks* 28: 68–90.

MISTRAL (2011). The sensor cloud the homeland security. http://www.mistralsolutions.com/hs-downloads/tech-briefs/nov11-article3.html (accessed March 2020).

NIEPMD (2014). National Institute for Empowerment of Persons with Multiple Disabilities (Manual), ISBN: 978-81-928032-1-0.

Peña-López, I. (2005). Itu Internet Report 2005: the Internet of Things, *Report no. 7*.

Popat, K.A. and Sharma, P. (2013). Wearable computer applications a future perspective. *International Journal of Engineering and Innovative Technology (IJEIT)* 3 (1): 213–217.

Raad, H. (2017). *The Wearable Technology Handbook*. Ohio: United Scholars Publications.

Raj, P., Raman, A.C., Nagaraj, D., and Duggirala, S. (2015). *High-Performance Big Data Analytics: The Solution Approaches and Systems*. London, UK: Springer-Verlag http://www.springer.com/in/book/9783319207438 (accessed July 2019).

Said, O. and Masud, M. (2013). Towards internet of things: survey and future vision. *International Journal of Computer Networks* 5 (1): 1–17.

Schnell-Davis, D.W. (2012). High tech casino advantage play: legislative approaches to the threat of predictive devices. *University of Nevada, Las Vegas Gaming Law Journal* 3: 299–346, Fall.

Sheng, Z., Yang, S., Yu, Y. et al. (2013). A survey on the IETF protocol suite for the internet of things: standards, challenges, and opportunities. *IEEE Wireless Communications* 20 (6): 91–98.

Thorp, E.O. (1969). Optimal gambling systems for favorable games. *Review of the International Statistical Institute* 37: 273–293.

Thorp, E.O. (1979). Systems for Roulette I. *Gambling Times* (January/February 1979).

Thorp, E.O. (1984). *The Mathematics of Gambling*. Secaucus, NJ: Lyle Stuart.

Vermesan, O., Friess, P., Guillemin, P. et al. (2011). Internet of things strategic research roadmap. In: *Internet of Things: Global Technological and Societal Trends*, vol. 1 (eds. O. Vermesan and P. Friess), 9–52. Aalborg, Denmark: River Publishers.

2

Applications

2.1 Introduction

As emerging technologies, IoT and wearables have given rise to a number of innovative applications and enabled the integration of smarter functionalities to outdated technologies. Not too long ago, such technology integrations were considered science fiction. This chapter covers the applications of IoT and wearables in various fields. It also provides an insight on the roles these applications could play in practice and discusses the challenges and key success factors for their adoption.

2.2 IoT and Wearable Technology Enabled Applications

2.2.1 Health care

IoT is indisputably transforming the healthcare sector by redefining the capacity of devices and people interactivity in delivering healthcare solutions. IoT has substantial applications in this sector that benefit patients, guardians, physicians, hospitals, and insurance firms.

In 2018, data from clinical trials of 357 head and neck cancer patients were presented at the American Society of Clinical Oncology (ASCO) annual meeting. The trials utilized a Bluetooth-based weight scale and blood pressure cuff, along with a smartphone app for symptom tracking which sends updates on symptoms and responses to treatment to their physicians every weekday. The patients who used this monitoring system experienced less illness and treatment side effects severity compared to the control group who maintained routine weekly physician visits with no additional monitoring at home. ASCO's president, Bruce E. Johnson, said that this technology "helped simplify care for both patients and their care

Fundamentals of IoT and Wearable Technology Design, First Edition. Haider Raad.

Published 2021 by John Wiley & Sons, Inc.

providers by enabling emerging side effects to be identified and addressed quickly and efficiently to ease the burden of treatment."

Thanks to their partnership with General Electric (GE) Healthcare and IoT-based software, known as AutoBed, Mt. Sinai Medical Center in New York City was able to effectively reduce the wait time by 50% for their emergency room patients. The system tracks occupancy among 1200 units and factors in 15 different metrics to assess the needs of individual patients.

Apart from monitoring patients' health, there are many other innovative uses where IoT devices are proven very useful in hospitals. IoT devices are used for tracking real-time locations of medical equipment such as wheelchairs, defibrillators, oxygen pumps, and other monitoring systems, in addition to asset management like pharmacy inventory control. Medical staff deployment at different locations can also be monitored and coordinated in real time. IoT-driven hygiene monitoring equipment is also used to help in preventing infections in patients.

In the health insurance sector, IoT can play a vital role in the underwriting and claims operations. Data gathered by health monitoring devices will enable the insurance companies to detect fraud claims and could bring transparency between the companies and customers in the processes of underwriting, pricing, claim handling, and risk management.

The first wearable hearing aid device was developed in 1938, which is considered as one of the early milestones in modern biomedical engineering. Nowadays, healthcare firms are realizing the potentials of wearable technology and IoT in treating some of the most common chronic diseases, such as, diabetes, congestive heart failure, hypertension, chronic obstructive pulmonary disease (COPD), and pain management. For example, a California-based medical tech company has developed a biosensor that monitors the patient's heart and respiratory rates and temperature, in addition to body posture. The device is also equipped with a fall detection capability and could be used in a home or a hospital setting. A wearable device developed by another healthcare firm is capable of continuously monitoring the glucose level in diabetic users. On the other hand, the FDA-approved Medtronic hybrid closed-loop system is capable of monitoring and adjusting glucose levels by automatically dispensing basal insulin, mimicking an actual function of a pancreas.

Another device, developed by iRhythm technologies Inc., is capable of detecting abnormal heart activity. The device is based on water-proof electrocardiogram patches which continuously collect heart data for two weeks. The collected data are then forwarded to the company's clinical app for processing and diagnosis.

In the pain management area, Quell is an FDA-approved wearable device aimed at reducing pain. The device utilizes an accelerometer which computes and assesses the user's activity level, and fires its pain reduction stimulation with a relevant intensities.

Another device available in the market is a pulse oximeter[1] which is developed for patients with asthma and COPD. The device can be used in either a hospital or home setting which allows remote and extended monitoring of the patient's oxygen level and heart rate.

There is also a wearable device that enables fetal monitoring for users with high body mass index (BMI). Another pregnancy device is designed to assist women in determining when they are ovulating and most fertile time for conception. The wearable device is inserted into the vagina where the biosensor monitors the basal body temperature.

Telemedicine is broadly defined as providing medical and healthcare services through telecommunication technologies. The first use of telemedicine dates back to the telephone invention in 1876, with medical consultations conveyed over the phone. Clearly, the recent breakthroughs in wearable technology and IoT have delivered medical care to virtually all corners of the world. With telemedicine, patients can access medical services that may not be available locally. Thus, transportation and geographic barriers are minimized. Moreover, it is widely acknowledged that cultural and social barriers may also prevent patients from seeking necessary mental health services in many cultures. Recent research findings have confirmed that telemedicine is very effective in diminishing such barriers. This area of telemedicine is known as tele-psychiatry.

Remote health care is now also available for pets. For example, the PetPace collar, an activity and wellness tracker designed for dogs and cats, can be linked to a network service where the pet's health parameters such as heart and respiration rates, temperature, and activity levels can be accessed by veterinary clinics.

2.2.2 Fitness and Well-being

The demand for fitness trackers that come packed with different types of sensing and wireless capabilities is growing at a rapid rate. Fitbit, Apple, Samsung, Jawbone, and Garmin are just a few brand names in today's market. These trackers measure fitness and well-being-related metrics such as the number of steps walked or climbed, heart rate, and sleep and stress patterns. Most trackers now also have the capability to determine the user's location.

A new trend started by many firms and organizations is the use of wearable devices to track the health and activity parameters of their employees as part of a well-being program. The aggregated data are then forwarded to their health insurance providers which, as an incentive, offer a reduced policy premium in return. According to one technology research and advisory firm, around 10000 companies across the world offered the use of fitness trackers to their employees in 2014.

---

1 An oximeter is a noninvasive device for monitoring a user's oxygen saturation levels.

It is worth noting that several studies agree that placing fitness trackers on the user's hip or foot rather than the wrist would result in more accurate readings. Even the most accurate fitness trackers in the market could overestimate step count in some scenarios and misinterpret exaggerated gestures as steps. This fact has motivated some wearable tech companies to design smart socks, bras, and undergarments. For example, one smart health tracking socks is capable of tracking the user's personal health metrics and offering higher accuracy in measuring steps, velocity, altitude, and burnt calories through their unique foot-landing and weight distribution techniques. The device consists of a cuff-shaped fitness tracker which magnetically connects to the product's running-friendly fabric. The product can also communicate with a smart phone app, keep logs of the user's activities, and guide them via audio cues during an activity.

Another wearable product in the area of well-being is a wristwatch that utilizes ultraviolet sensors to track the levels of sunlight exposure received by the body. The data are then visualized via LED lights that start to flicker when the user's ultra violet exposure is within the dangerous level. Such device would be very practical in countries with high skin cancer rates (i.e., Australia, New Zealand, Argentina, Denmark, and parts of USA) due to excessive exposure to ultraviolet energy.

One of the main reasons IoT technology has had such an impact on the fitness industry is the visibility it offers. Regardless of the user's exercise goals, their primary objective is to improve in some area and be able to quantify such improvements. Through continuous data collection, analysis, and visualization, IoT provides the users with unprecedented visibility to track personal growth. For example, IoT-driven smart home bikes and elliptical workout machines that feature streaming workouts, cycling classes, and other digitally connected features are on the rise. Such equipment offers flexibility for people with no time to go to a gym and the sense of participating in real classes. Such demand has also triggered software enterprises, such as Kaa, to develop IoT platforms that deliver production-ready capabilities into smart sport and fitness products. Such platforms allow manufacturers to automatically aggregate and analyze data from virtually any sensors, fitness trackers, and smart sports equipment, and then visualize it on equipment displays and/or mobile devices.

2.2.3 Sports

Today, IoT and wearable technology play a vital role in sports through athlete development and safety, and fan engagement and experience. Organizations are investing billions of dollars on smart stadiums where IoT is used to improve digital engagement and in-arena experience. Fans can have an immersive experience with their favorite teams and athletes like never before.

In the area of player development, IoT is transforming the way coaches coordinate training, manage players, and address essential situations in every game. Integrating advanced game analytics with sensors, coaches can easily access vast amounts of processed data to obtain players' efficiency and performance metrics, in addition to opponent shortcomings to develop a more educated in-game strategy. Moreover, embedded sensors and microchips offer sport physicians and therapists real-time health tracking which provides a holistic view of the athlete's state, allowing them to make a more informed decision for the athlete's longevity and health status.

For example, Adidas is working with professional soccer teams in parts of the United States to monitor the heart rate and other metrics of players using its miCoach wearable technology. The aggregated data are analyzed by coaches to track the athletes' performance and have the best decision on scheduling breaks to minimize the risk of injury. Other sensing devices are worn underneath the athletes' garments and used to monitor other key parameters including velocity, orientation, acceleration, blood pressure, and heart rate, which then sent to the coach's console.

Another sports wearable device is designed for alpine sports with onboard processing power, sensory, and networking capabilities comparable to that of a smartphone. The wearable gear is intended for skiers and snowboarders to stay connected. It is capable of displaying the user's speed and altitude, route maps, and social network profiles and locations of other skiers in the resort. It simply gives the user the power of a smartphone integrated into their field of vision, hands-free.

There are other wearable products available for hobbyists which are capable of tracking heart rate, muscle tension, and breathing patterns, and alert the users via text or push notification. Other products include smart sport apparatus such as a baseball that can detect and display the speed, spin rate, and pitching trajectory when thrown.

2.2.4 Entertainment and Gaming

Technologies like virtual reality (VR), augmented reality (AR), and haptics[2] are redefining the way we experience and make movies, music, and video games. For example, Oculus Rift, a virtual reality gadget, is being used to create immersive movies that are unparalleled to what we have experienced before.

2 **Haptics**, also known as kinesthetic communication, is aimed at reconstructing the sense of touch by applying forces, vibrations, or motions to the user through motors and other actuators for the purpose of interacting with computer applications.

Wearable devices with haptic feedback and gesture control are now used to make innovative forms of music. For instance, Imogen Heap, an English singer and composer, introduced smart gloves which utilize gestures and motions to create digital music.

Some theme parks around the world started to employ wearable technology to enhance the entertainment experience of tourists. For example, Disney introduced a smart wristband to help tourists navigate their theme parks. The tourist information is linked to a database by the smart band which also serves as an admission ticket, hotel key, and credit card. Tourists are able to schedule their visits to each theme park and preorder food without the need to wait in the extremely long lines.

Market research reports that the global market for the wearable interfaces of virtual reality is estimated to reach 1 billion US dollar by 2020. It is also expected that the use of wearables such as wristbands, earbuds, and eyewear will increase exponentially in gaming platforms as control devices for the virtual reality and biometric gaming market due to their potential to accurately track the gamer's movements. According to one company that develops biometric technology for wearables, an exciting potential application in immersive gaming may include action games that require gamers to hold their breath while the character is under water. Another application would be for the user's heart rate to directly affect their accuracy in a shooting-based game.

2.2.5 Pets

In 2017, American Humane organization reported that around 10 million cats and dogs are either stolen or lost in the United States every year. It is also estimated that about 60% of cats and 56% of dogs in the United States are either obese or overweight. Such statistics have driven innovators to find IoT and wearable solutions dedicated to pets.

According to a recent market report, the global revenue of pet wearables market was $1.4 billion in 2018, and is expected to reach 2.36 billion US dollar by 2022. The growth in pet ownership and expenditure is expected to push product demand over the forecast period. Moreover, the increased awareness toward pet health is driving substantial investment for research and development of even more advanced wearable products for pets. One example of a pet wearable is an on-collar tracker that is able of collecting data of the pet's physical activity through a smartphone app and keeping records of its behavior. Such behavioral data allow pet owners to determine whether the pet is potentially ill or has some condition; it also allows owners to track the location of their pets.

While most IoT products emphasize on the health and security for pets, there are other areas in which IoT can be helpful for pet owners, such as pet toys and feeders. Feeders can help automate the precise portion that a pet should have while also dispensing food when the owner is away. Some apps even have the capability to remotely unlock and lock home doors to allow access for dog walkers and pet sitters.

2.2.6 Military and Public Safety

Internet of Battlefield Things (IoBT) or Internet of Military Things (IoMT) uses sensing, communication, and computing devices embedded within the soldiers' combat suits, helmets, weapons, and other equipment to provide them with additional sensory perception, situational understanding, and a better response time.

Wearable technology has become a fundamental component of the connected soldier system that offers a tactical edge, tracking, and improved safety. The research and development arena is dedicating greater efforts in this field to advance lightweight electronics and miniaturized antennas, and to produce more effective radio communication systems with an eye toward improving soldier mobility and system scalability. Such systems involve a variety of physiological sensors that monitor heart rate, breathing patterns, body temperature, and blast effects. The data would be available for real-time wireless transmission to headquarters or for immediate analysis.

In addition to worn devices, the US Army Research Laboratory (ARL), academia, and industry are collaborating to investigate how other stationary and mobile devices' connectivity and infrastructure can be utilized to improve military operations. ARL has been testing various protocols and technologies to bring devices and sensors together in a cohesive network. In addition to soldiers, ships, planes, tanks, drones, and weapons are all entities of such network.

On the other hand, many law enforcement departments are already using wearable headsets which include a camera, an ear phone, and a microphone, used when on duty. For instance, London Metropolitan Police had deployed wearable systems aiming at regaining the public trust after the department's involvement in controversial cases following the killing incidents that led to the London riots in 2011. In spite of the escalated privacy concerns, the plan seems to have exhibited success in documenting offense evidences which resulted in a sped up justice process. The author predicts that virtually every police officer in the United States will be equipped with a wearable system which could include a camera and recording unit by 2025. They will most likely be equipped with a wrist or eyewear for accessing information, hands-free.

Recent studies report that wearable technology can be crucial in field communications and would enhance situational awareness in civil defense and public safety applications. This would promote multitasking and ultimately results in an improved decision making process. One example is represented by a wearable device developed by Vienna University of Technology which was designed to aggregate data captured by firefighters. The device is capable of mapping the firefighters' surroundings on a given site by a color-coded representation. The temperature mapping is then used to identify the location of people through smoke and help in deciding whether specific rooms are safe to gain access to. This is accomplished by wearable cameras and sensors integrated within the firefighter's helmet.

2.2.7 Travel and Tourism

The travel and tourism industry has not been shy about investing in IoT and wearable technology. The sector spent about $128.9 million on these technologies in 2015. IoT is now being used to streamline the end operations and maximize operational efficiency of hotels, airlines, and travel agencies by interconnecting smart devices, systems, and processes. For example, hotels and airlines can track supply chains more effectively through sensor-enabled cargos, which enables them to account for any contingencies and prevent service disruptions to travelers.

Many studies report that wearable and mobile technologies are starting to play a vital role in many aspects of the travel and tourism field. For examples, several deployments of Google Glass in the airline business have been recently reported. Also, in 2014, Virgin Atlantic staff at Heathrow Airport has used the Google Glass to provide an enhanced customer service. The Glass enabled the airlines' employees to offer weather information and language translations to their customers.

The hospitality industry is another sector adopting wearable technology. For instance, Starwood Hotels developed a technology that enables their preferred guests to use virtual room keys instead of physical ones. It also offers directions to the hotel and allows access to their reservations and star point balances.

2.2.8 Aerospace

Until recent times, astronauts relied on printed instruction manuals in case of emergency or system error. Such issues force the crew to call the ground station for guidance. However, telecommunication becomes impractical the farther a spacecraft is away from earth. For example, it could take up to 25 minutes for a message to travel from Mars to Earth. To overcome such problems, the U.S. National Aeronautics and Space Administration is developing smart glasses for astronauts that can guide them through a repair process or conducting an experiment in outer space, hands-free.

IoT is revolutionizing the aerospace industry, both on the ground and in the air. Real-time analytics via IoT are already pushing improvements in quality and manufacturing efficiency in this sector. For example, IoT-enabled power meters can provide information on energy usage in aircraft production, which could lead to significant cost reductions and a more sustainable operation. According to Airbus, advanced analytics algorithms analyze the energy usage and suggest energy-saving measures which could result in 20% cost saving.

IoT can also offer a more in-depth insight into how an entire assembly line is operating. In one aircraft factory, data from machines and conveyors are fed into a live visual hub to enable supervisors to track operations in real time, as well as implement highly accurate simulations to obtain the most optimized ways of improving operations.

2.2.9 Education

Many educators are starting to realize that the emerging digital technologies could offer an opportunity to enhance the learning experience instead of being a distraction. In fact, many studies are confirming the advantageous potential of using IoT and wearable technology as pedagogical tools.

For more than three decades, Microsoft's PowerPoint has served educators in almost every discipline as an indispensable illustrative tool. However, postmillennial generations will most likely enjoy a more immersive classroom and learning experience that go beyond a simple slideshow. In response to this, hundreds of classrooms across the world have already started to deploy wearable and IoT platforms to transform the learning experience of students. For example, Google Expeditions, an educational initiative introduced by Google, uses folded cardboard with a pair of specially designed lenses to turn a smartphone into a virtual reality headset. This gadget lets students have an immersive visual learning experience. The gadget serves as a great tool for active and hands-on learning where a user can virtually explore places such as the Great Wall of China, the Grand Canyon in the United States, or Rome in Italy.

"The creativity we have seen from teachers, and the engagement from students, has been incredible.", as reported by Google's product manager for Expeditions. This is one example where technology in education, if used properly, can enhance the student's creativity and learning experience without distractions.

IoT could also help schools improve the safety of their campuses, keep track of essential resources, and improve access to information. Connected devices can be used to monitor students, staff, and equipment at a reduced operating cost.

Last but not least, using IoT-enabled devices is an effective way to provide educational assistance to disabled students. Hearing-impaired students may utilize a system of smart gloves and a tablet or a smartphone to translate from sign language to verbal speech, and vice versa.

2.2.10 Fashion

Today, when we discuss the topic of wearable technology, the first thing that comes to mind is the plain-looking smartwatches and fitness trackers which, from the fashion critics' perspective, are still lacking in "style." As wearable technology is marching toward becoming mainstream, developers realize that collaborating with fashion designers is crucial to create trendy products people would actually want to buy.

One of the first wearable industry–fashion collaborations was between Martian and Guess to produce the Guess Connect Smart Watch. Tory Burch has also designed various accessory that go with Fitbit fitness tracker. Swarovski is also collaborating with Misfit trackers on the Misfit Shine fitness wristband which can be controlled by a large Swarovski crystal. More interestingly, there is a solar powered version, which utilizes light refracted by the crystal to power the tracker.

Tag Heuer's first smartwatch released in 2015 was one of the most significant collaborations between a watchmaker and a high tech company. The Tag Heuer Connected was a collaborative work of Intel and Google, running Android Wear embodied by custom Tag Heuer skins.

The fashion guru Ralph Lauren has also entered the scene with the Polo Tech Shirt. The smart shirt has sensors woven into the shirt fabric which are capable of tracking metrics such as heart rate, pressure, temperature, and breathing patterns. Needless to say, such smart shirts come with a hefty price tag.

2.2.11 Business, Retail, and Logistics

According to a recent survey conducted by Forbes which involved 700 executives, 60% of enterprises are using IoT to expand or transform new lines of business, with 63% of these enterprises already supplying their customers with new or updated services utilizing IoT-enabled capabilities. Such capabilities provide businesses with preventive maintenance, automated product updates, in addition to inventory tracking. Sensors located in production systems, assembly lines, warehouses, and vehicles generate data that help management understand how operations are moving along and to gain greater insights into the productivity and performance of their systems and processes, which ultimately offer opportunities for innovation and growth.

On the other hand, giant firms and medium-sized businesses are taking initiatives in harnessing the benefits of wearable technology in the workplace. The promise of data captured from monitoring staff movements to improve productivity and efficiency might be very tempting to management, but it comes with a pressing issue: privacy concerns, which will be discussed in Chapter 9 of this book.

Retailers have worked with RFID for decades. However, the opportunities to improve operations and offer personalized and immersive experiences for customers would not be possible without IoT. For example, the jewelry maker Swarovski is using virtual reality to create new forms of customer experiences, such as watching the process of product creation, or making a purchase at the store, all while wearing a virtual reality headset.

Smart mirrors, for example, are being used by high-end retailer Neiman Marcus in New York City. Smart mirrors allow customers to search for other sizes or colors when trying a product, suggest matching items, or even show how a product might fit without physically trying them. By providing immersive experience, supplemental information, and a fluid customer experience, Neiman's smart mirror was able to expand sales.

It is obvious that with several IoT applications in areas such as dynamic pricing, smart shelves, and inventory management it is possible to accomplish multiple potential outcomes. IoT is also capable of facilitating the ongoing optimization of business processes and even shaping the employee engagement and performance. In some industries, IoT is already being used in supply chains to autonomously execute transactions when certain conditions are met.

2.2.12 Industry

In industrial settings, the potential market for wearable technology solutions is expected to exceed that of the general smart living consumer market. Firms in the field of services have already witnessed the impact of wearable technology, with technicians and engineers wearing camera-based headsets while on field jobs. For example, Vuzix produces a variety of glasses and headsets that offer innovative solutions for warehouse management systems. Fujitsu, on the other hand, is focusing on a smart glove product designed for industrial maintenance and on-site operations. The glove is integrated with a Near Field Communication (NFC) tag reader and features a gesture-driven input controller.

2.2.12.1 The Industrial Internet of Things (IIoT)

The application of the IoT in the industrial sector commonly referred to as IIoT is becoming extremely pervasive and is revolutionizing the manufacturing and industrial processes by enabling more efficient acquisition and accessibility of data, at greater speeds. Whether by enabling analytics to detect erosion inside a refinery pipe, offering real-time production data to pinpoint untapped potential capacity in a plant, predicting maintenance, improving safety, or accelerating new product development by feeding back operative data into the product design cycle, IIoT is driving powerful outcomes.

IIoT aims at bringing together machines, advanced analytics, workers, and managers. In essence, it is a network of industrial units connected by communications technologies that result in systems that are capable of monitoring, gathering, analyzing, and delivering invaluable insights that can help drive smarter and faster business decisions.

IIoT tool-based advanced analytics solutions help industries increase asset reliability and availability while minimizing maintenance costs and preventing operational risks. For example, one tool is designed to increase field service efficiency and improve customer experience. The tool enables dispatchers to schedule and dispatch servicing jobs, while guiding technicians to accomplish an optimized service delivery by providing expert instructions, equipment data, and customer information. Another tool developed by General Electric (GE) is aimed at helping operators take the right actions to take every time utilizing model-based high performance Human–Machine Interface (HMI) for faster response and development.

Such IIoT solutions are helping industrial organizations drive substantial gains in productivity, availability, and longevity.

2.2.13 Home Automation and Smart Living

Home automation may encompass centralized control of lighting, air conditioning systems, appliances, security units, and other devices to provide improved convenience, comfort, and energy efficiency. The concept of home automation has been around for decades, and products have been available to the consumer for a long time.

Arguably, the first generation of smart homes had almost nothing to do with intelligence and was more about remote control, often through a smartphone or a computer, with only a slight automation. Today, however, a good number of modern living spaces have blinds that can adjust its angle to maintain a certain brightness or a thermostat that can adapt to a temperature setting learned from observing a user's habits and determine consumption patterns using complex algorithms. These insights then help the users personalize their experience at a microlevel. This indeed is a "smart" home. For example, some smart thermostats available in the market use sensor data and special algorithms to automatically adjust schedules, and monitor the user's location in real time to turn the air conditioner on and off accordingly. Moreover, smart fridges are capable of preordering milk and egg, and check the expiration dates on the products to help the user optimize their shopping list.

The market of IoT tools and services for smart home is very broad and diverse. Some manufacturers focus on specific areas of the household environment, for example, lighting, and temperature regulation. Others develop complete hubs for a smart home that are able to connect and communicate with other smart devices, such as Amazon Echo or Google Home virtual assistants.

IoT-driven smart home systems enable transparency to the user's household which results in optimized utility spending. Using insights generated from smart

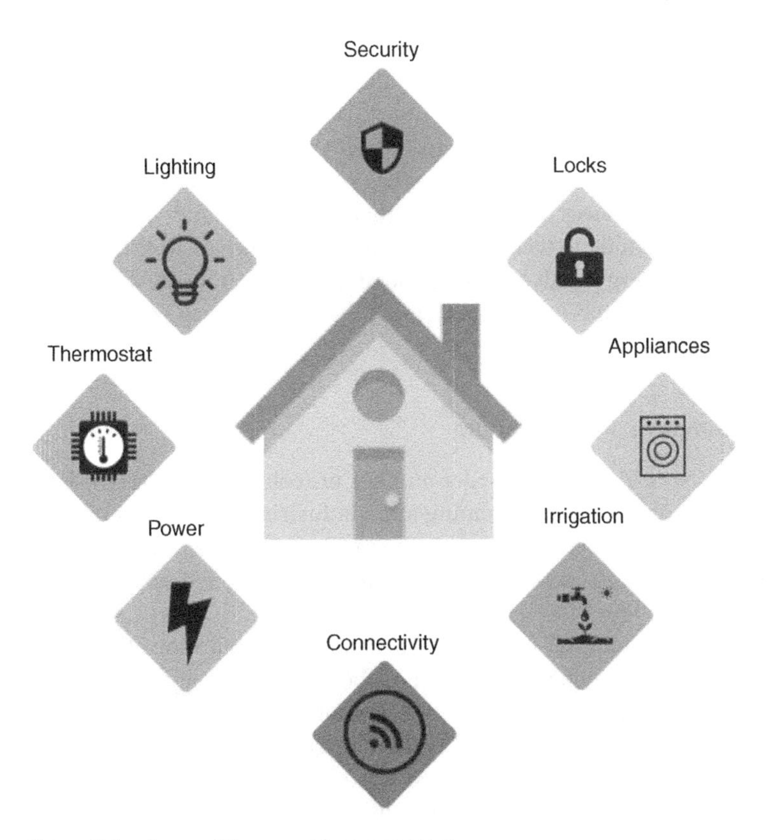

Figure 2.1 Areas of the smart home concept.

home devices on electricity, water, and gas consumption, users can easily identify the energy-wasting points and habits and adjust usage accordingly (Figure 2.1).

2.2.14 Smart Grids

Smart Grid technology is arguably one of the greatest implementations of IoT in the field of energy, which could tremendously help with energy resources conservation.

According to the U.S. Department of Energy (DoE), current power outages and interruptions cost Americans at least $150 billion each year. As the world's population continues to grow, the classical grids will not be able to keep up with the rising demands. Smart Grids are designed to lower costs through IoT-driven monitoring and source rerouting once a power failure is detected.

The Smart Grid is part of an IoT framework, which can be used to monitor and manage everything from lighting, traffic signals, parking spaces, and early detection of power influxes resulting from seismic activities or extreme weather. This is enabled by a network of transmission lines, smart meters, sensors, substations, and data analytics.

Smart Grid technologies contribute to efficient energy management solutions and are integral to establishing a smart city. The two-way communication between connected devices and systems that can sense and dynamically respond to user demands is what makes smart grids superior to the existing framework. Smart energy analytics can gather data on power loads, water flow, pressure, temperature, and other parameters to help users keep track of their consumption habits.

2.2.15 Environment and Agriculture

The agriculture sector has witnessed a number of technological transformations over the last few decades, becoming more industrialized and technology-driven. By using various smart machinery and instrumentation, farmers have attained better livestock and crop management, which led to improved productivity.

The term "Smart agriculture" is often used to denote the application of IoT solutions in agriculture. Although it is not as popular as consumer connected devices, the adoption of IoT solutions in agriculture is constantly growing. According to a recent market study, the global smart agriculture market size is expected to reach $15.3 billion by 2025, compared to around $5 billion in 2016.

Crop management is one area IoT is used in agriculture. Sensor-enabled stations are placed in the field to collect data specific to crop from temperature, humidity, and precipitation to plant water potential and overall crop health. Thus, crop growth and any anomalies can be monitored to effectively prevent any diseases or infestations.

Cattle monitoring and management are another application where sensors are attached to the farm animals to monitor their health and performance. For example, one product available in the farming market today uses smart collar tags that are capable of delivering temperature, health, activity, and nutrition insights on each individual cow as well as cumulative information about the herd.

A more advanced application of IoT-based agriculture can be represented by the farm productivity management system which includes a number of sensors and communication units installed on the field premises, in addition to a user interface with analytical capabilities. Such systems allow the user to remotely monitor the farm and streamline its business operations. Other IoT-driven agriculture use cases include vehicle tracking, greenhouse automation, and storage management.

It is also anticipated that IoT will have a positive impact on the environment. Gadgets that give us insights and help people conserve energy, water, and resources, or sensors that measure radiation, air and water quality, or detect hazardous chemicals, already exist in the market, but there are also more innovative ways IoT is used, such as in preservation of biodiversity, prevention of deforestation, and poaching. For example, deforestation accounts for 15% of global emissions from carbon. Some organizations have initiated IoT projects aiming at preventing further deforestation by stopping illegal activities through smart sensors attached to trees that allow them to remotely monitor and detect illegal logging and poaching.

2.2.16 Novel and Unusual Applications

The reader might not think that wearables have found an intuitive use in baby diapers, but Huggies, a leading diaper manufacturer has developed a moisture sensor tag that can be attached to the outside of a baby's diaper. The tag links to the parent's smartphone and sends a notification when the diaper needs to be changed. While the device is mainly aimed at helping parents, there is also a marketing catch to it.

A smart fabric startup is utilizing galvanic skin response, with sensors placed on the user's hands to analyze signals that determine their mood. The company's smart shirt, the Mood Sweater, has a collar of LED lights which changes color to reflect the wearer's mood. Another company has developed a printed electro-biochemical sensor which gives the user information about blood and biomarkers such as pH and sodium levels by applying a temporary tattoo on to the user's skin.

Pollution poses significant health risks, and it is a major issue in many urban cities. The Climate Dress developed by Diffus might be the first wearable technology aimed at raising awareness and alerting users to avoid places with high levels of pollution via sensors that measure the Carbon Dioxide concentration in the air.

The Hug Shirt designed by CuteCircuit allows people to send hugs to loved ones over distance. The shirt is equipped with sensors and actuators that garner and emulate the strength, duration, and location of the hug. Moreover, the skin warmth and the heart rate of the sender can also be reproduced.

Statistics report that one in eight women will develop breast cancer during their lifetime. Regular self-examination and regular physician visits are of paramount importance since early detection can be a lifesaver. However, a large percentage of women either forget about the self-examination or are unaware of the consequences. This fact triggered Nestle` to create the Tweeting Bra which is activated by a concealed mechanism under the bra's hook. Every time the bra is taken off, it notifies the user's smartphone, which in turn generates a tweet. The tweet, for example, will read: "Kelly has just unhooked her bra. When you do the same, don't forget about your self-exam."

The Aurora Dreamband is the first smart wearable gadget designed to enhance the awareness and perception of rapid eye movement (REM) dreaming. Aurora analyzes the user's brainwaves and body movements using an accelerometer, EEG, ECG, EMG, and EOG sensors.

There are also the connected egg tray that notifies the user when they're about to be out of eggs, and smart water bottles that could pair with an app over Bluetooth, then constantly remind the user to stay properly hydrated!

2.3 Smart Cities

Cities and metropolitan areas are vital to global economic development. The concentration of people and businesses in cities promotes ideal conditions that give rise to new industries and technological innovation. In the United States, urban metropolitan areas are home to 85.7% of the country's population, 87.7% of total employment, 87.9% of total income, and 98.9% of the increase in Gross Domestic Product (GDP), which is a monetary measure of the market value of goods and services. However, cities are under a rising pressure due to the exponential growth in urban population which is leading to drained infrastructure and resources. For instance, most densely populated cities suffer from traffic congestion, which in turn gives rise to air pollution and health issues. Other challenges facing urban areas include public safety and the difficulties experienced by an aging society. Cities need to make determined efforts to deal with such issues in order to maintain economic success and remain attractive to citizens. To achieve this, many cities are investing in smart city projects to improve efficiency, handle complexity, and enhance the citizens' quality of life.

The term "smart city" was conceived toward the end of the past century. It is based on the implementation of information and communication technologies in applications dedicated for future cities and their development. Smart cities promote progressive social and technological innovations and link existing infrastructures with an eye toward optimizing the quality of life. They incorporate novel energy, traffic, and transportation concepts with minimal adverse effects on the environment. They also focus on new forms of governance and public involvement, and finding remedies to the current global challenges, such as climate change and shortage of resources.

Functional areas of smart cities include but are not limited to:

- Traffic and transportation which include smart ticketing, smart parking, and autonomous transport systems.
- Energy and resources which include smart grid, environmental sensors, and irrigation management.
- Urban infrastructure which includes smart street lighting, smart buildings, and waste management systems.
- Smart Governance which includes consolidated services platforms and reporting systems.
- Safety and security which include integrated video surveillance and predictive analytics.
- Health care which includes telemedicine and remote patient monitoring.

Figure 2.2 depicts the major areas where a smart city can incorporate.

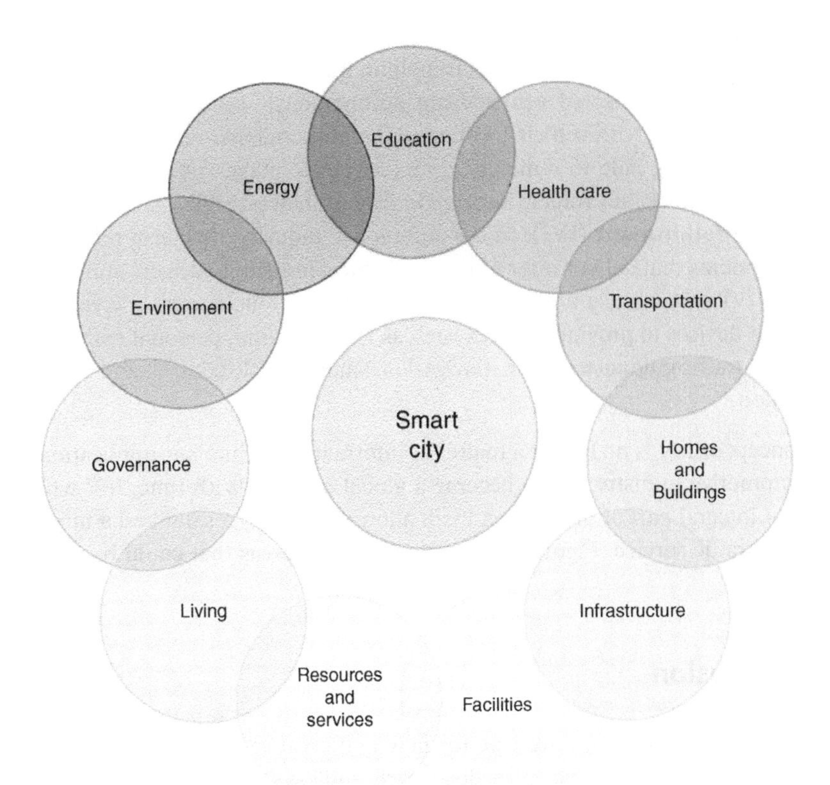

Figure 2.2 Functional areas of smart cities.

2.4 Internet of Vehicles (IoV)

Internet of Vehicles (IoV) is envisioned to serve as a key sensing and processing platform for data in smart transportation systems. A vehicle will be equipped with a sensor network, acquiring information from the roads, other vehicles, and the driver and employ it for safe navigation and traffic management. IoV comprises vehicles that communicate with each other (vehicle-to-vehicle (V2V)), vehicle-to-road (V2R), vehicle-to-human (V2H), and vehicle-to-sensor (V2S) interconnectivities, forming an intelligent network of objects and users.

In addition to autonomous (self-driving) cars, IoV has various functional applications which include the following:

Driving Safety: This is often portrayed by a collision avoidance system that employs various sensor technologies to detect impending collisions and immediately alerts the driver.

Traffic Control: Aims at bringing improvement to traffic congestion management, transport and logistics, and metropolitan traffic.

Crash Response: Connected vehicles can autonomously send real-time data about a crash along with vehicle location to first respondents.

Convenience: Being able to remotely access a vehicle makes services such as remote door unlock and vehicle location in case of theft possible.

In-Vehicle Infotainment (IVI): In the automotive industry, this term refers to vehicle systems that deliver information and entertainment to drivers and passengers. IVI systems use audio and video interfaces, touchscreens, keypads, and other devices to provide services such as toll collection, personal communications, traffic guidance system, navigation, smart vehicle control, and crash prevention.

The concept of IoV is no longer a matter of information technology application in the automotive industry; it has become a global interest. With time, IoV will become an integral part of our life, and will allow us to enjoy a safer and a more convenient traffic service. Figure 2.3 depicts the major areas that could be enabled by IoV.

2.5 Conclusion

IoT and wearable technology are ripe for new and creative ideas to add to the applications already in use. They provide a nearly endless supply of opportunities to interconnect our devices and equipment. When it comes to innovation, this field is wide open, and such connectedness will substantially reshape our world in ways we can barely imagine.

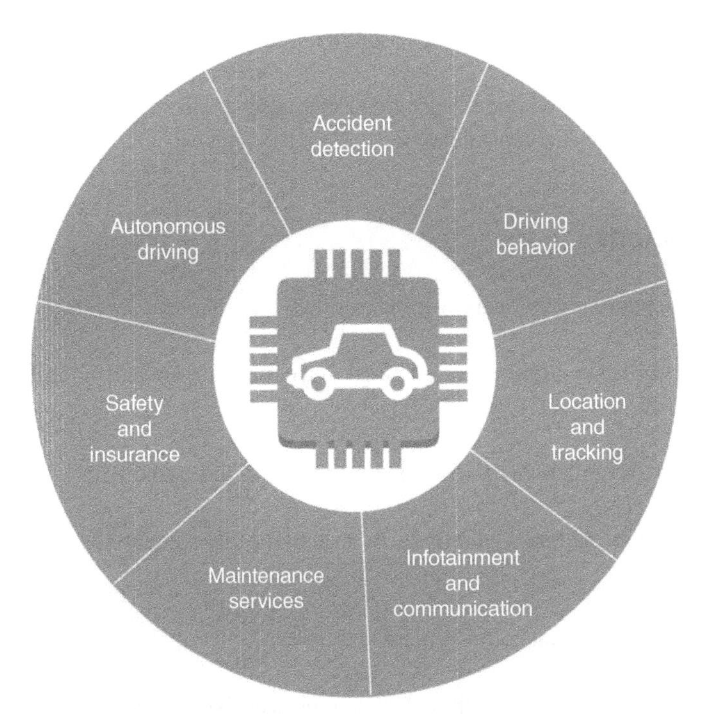

Figure 2.3 The major areas of IoV.

Problems

1. Can you think of more applications (other than the ones listed in this chapter) that could benefit from IoT and wearables?

2. Create a novel scenario where drivers and/or pedestrians could benefit from IoV.

3. Create a novel scenario where governments could benefit from IoV.

4. Create a scenario where home automation is utilized in the field of safety.

5. Could you think of more potential applications of IIoT?

6. List five unusual applications where IoT and wearables are utilized. Keep efficiency and practicality in mind, and make sure that no products exist that support such applications (through an internet search).

7 List 10 applications where wearables are used in health care.

8 Write a one page scenario where at least ten of the applications mentioned in this chapter are utilized in a typical day.

Interview Questions

1 Talk about some popular platforms of Industrial IoT.

2 What impacts will the Internet of Things have on infrastructure and Smart Cities?

3 What are the key differences between Consumer IoT and Industrial IoT?

4 What are the different sectors where the IoT can actually add value to the current processes?

5 What are the main areas that could benefit from IoV?

6 What is a smart city and why do you think IoT is a crucial enabler of this concept?

7 What are the main areas in home automation?

8 Which IoT and wearable sector has the most impact/revenue share?

Further Reading

Abbasi, M.A.B., Nikolaou, S.S., Antoniades, M.A. et al. (2017). Compact EBG-backed planar monopole for BAN wearable applications. *IEEE Transactions on Antennas and Propagation* 65 (2): 453–463.

Adjih, C., Baccelli, E., Fleury, E. et al. (2015). FIT IoT-LAB: A large scale open experimental IoT testbed. *IEEE World Forum on Internet of Things (IEEE WF-IoT)*, Milan, Italy (December 2015).

Aho, A.V., Sethi, R., and Ullman, J.D. (1986). *Compilers, Principles, Techniques*. Boston, MA: Addison Wesley.

Alrawi, O., Lever, C., Antonakakis, M., and Monrose, F. (2019). SoK: security evaluation of home-based IoT deployments. *IEEE Security and Privacy (SP)*, NY, USA.

Amazon AWS IoT (2018). The internet of things with AWS. https://aws.amazon.com/iot/ (accessed February 2020).

Android API (2018). Android sensor API documentation. https://developer.android.com/guide/topics/sensors/sensors_overview.html (accessed February 2020).

Android Monkey (2018). UI/application exerciser. https://developer.android.com/studio/test/monkey (accessed February 2020).

Android Things (2018). Android things official apps. https://github.com/androidthings (accessed February 2020).

ARCEP (2016). Livre blanc – Préparer la révolution de l'Internet des objets – ARCEP | IoT, utorité de Régulation des Communications Électroniques et des Postes, Paris.

AT&T INC (2017). Annual report. https://investors.att.com/~/media/Files/A/ATT-IR/financial-reports/annualreports/2017/complete-2017-annual-report.pdf

Automobile (2017). Automobile mag, the big data boom. *Automobile Magazine* (10 October 2017), http://www.automobilemag.com/news/the-big-data-boom/

Baumann, L.M. (2016). The story of wearable technology: a framing analysis. MA Thesis. Virginia Polytechnic Institute and State University, Blacksburg, VA. 4.

BBC (2016). Walk with the world's biggest dinosaur in virtual reality. http://www.bbc.com/earth/story/20160219-attenborough-and-the-giant-dinosaur-virtual-reality-360 (accessed 25 June 2016).

BEREC (2016). BEREC Report on Enabling the Internet of Things, *Report no. 39.* http://berec.europa.eu/eng/document_register/subject_matter/berec/reports/5755-berec-reporton-enabling-the-internet-of-things.

Berkay Celik, Z., McDaniel, P., and Tan, G. (2018). Dynamic enforcement of security and safety policy in commodity IoT. arXiv preprint (2018).

Berkay Celik, Z., McDaniel, P., and Tan, G. (2018). Soteria: automated IoT safety and security analysis. *USENIX Annual Technical Conference (USENIX ATC).* Boston, MA.

Boateng, G., Batsis, J.A., Halter, R., and Kotz, D. (2017). ActivityAware: an app for real-time daily activity level monitoring on the Amulet wrist-worn device. *IEEE International Conference on Pervasive Computing and Communications Workshops (PerCom Workshops).* HI, USA: IEEE, pp. 431–435.

Bower, M. (2010). Affordance analysis–matching learning tasks with learning technologies. *Educational Media International* 45 (1): 3–15.

Bower, M. and Sturman, D. (2015). What are the educational affordances of wearable technologies? *Computers & Education* 88: 343–353. Canberra.

Celik, B. (2018). Peek-a-Boo: I see your smart home activities, even encrypted! arXiv preprint arXiv:1808.02741 26.

Chan, M., Estève, D., Fourniols, J.-Y. et al. (2012). Smart wearable systems: current status and future challenges. *Artificial Intelligence in Medicine* 56: 137–156.

Chatterjee, A., Aceves, A., Dungca, R. et al. (2016). Classification of wearable computing: A survey of electronic assistive technology and future design. *Proceedings of the 2016 Second International Conference on Research in*

Computational Intelligence and Communication Networks (ICRCICN), Kolkata, India (23–25 September 2016), pp. 22–27.

Chi, H., Zeng, Q., Du, X., and Yu, J. (2018). Cross-app threats in smart homes: Categorization, detection and handling. arXiv preprint arXiv:1808.02125 (2018).

Chiauzzi, E., Rodarte, C., and DasMahapatra, P. (2015). Patient-centered activity monitoring in the self-management of chronic health conditions. *BMC Medicine* 13: 1–6.

Choudhary, S.R., Gorla, A., and Orso, A. (2015). Automated test input generation for Android: Are we there yet? arXiv preprint arXiv:1503.07217 (2015).

CISCO (2017). Cisco Visual Networking Index: Global Mobile Data Traffic Forecast Update 2016-2021, *Report no. 738429*. https://www.cisco.com/c/en/us/solutions/collateral/service-provider/visualnetworking-index-vni/mobile-white-paper-c11-520862.pdf.

Clarke, E.M. and Emerson, E.A. (1981). Design and synthesis of synchronization skeletons using branching time temporal logic. *Workshop on Logic of Programs*, NY, USA.

Clause, J., Li, W., and Orso, A. (2007). Dytan: a generic dynamic taint analysis framework. *ACM Software Testing and Analysis*, London UK.

Coffman, T. and Klinger, M.B. (2015). Google Glass: using wearable technologies to enhance teaching and learning. Paper presented at the Society for Information Technology & Teacher Education International Conference, Las Vegas (2–6 March 2015).

Comitz, P. and Kersch, A. (2016). Aviation analytics and the internet of things. *Integrated Communications Navigation and Surveillance (ICNS)*, Herndon, VA, USA, 2016.

Dogo, E.M., Akogbe, A.M., Folorunso, T.A. et al. (2014). Development of feedback mechanism for microcontroller based SMS electronic strolling message display board. *African Journal of Computer & ICTs* 7 (4): 59–68.

Dumanli, S. (2015). Challenges of wearable antenna design. Presented at ARMMS Conference, Oxford, UK.

eeNews Automotive (2017). Automotive news, Fiat Chrysler joins autonomous driving platform from BMW/Intel/Mobileye | EETE Automotive. http://www.eenewsautomotive.com/news/fiat-chrysler-joinsautonomous-driving-platform-bmw-intel-mobileye.

Google Cloud Platform (2017). Designing a connected vehicle platform on cloud IoT core solutions, Google Cloud Platform. https://cloud.google.com/solutions/designingconnected-vehicle-platform.

GSMA (2017). Mobile IoT - Internet of Things. https://www.gsma.com/iot/mobile-iot-executivesummary/ (accessed March 2020).

Gubbi, J. (2013). Internet of Things (IoT): a vision, architectural elements, and future directions. *Future Generation Computer Systems* 29 (7): 1645–1660.

Hazarika, P. (2016). Implementation of smart safety helmet for coal mine workers. *Proceedings of the 1st IEEE International Conference on Power Electronics, Intelligent Control and Energy Systems*, Delhi, India (4–6 July 2016), pp. 1–3.

Heintzman, N.D. (2016). A digital ecosystem of diabetes data and technology: services, systems, and tools enabled by wearables, sensors, and apps. *Journal of Diabetes Science and Technology* 10: 35–41.

Stanford Medicine (2020). The Autism Glass Project at Stanford Medicine. http://autismglass.stanford.edu/ (accessed March 2020).

Luckerson, V. (2015). Google will stop selling glass next week. *Time Magazine* (15 January 2015). http://time.com/3669927/google-glass-explorer-program-ends/.

Elgan, M. (2016). Why a smart contact lens is the ultimate wearable. *Computer World* (9 May 2016). http://www.computerworld.com/article/3066870/wearables/.

Google.com (2020). Google developers, platform overview. https://developers.google.com/glass/develop/ (accessed September 2020).

Pierce, D. (2015). iPhone killer: The secret history of the Apple watch, Wired.com. https://www.wired.com/2015/04/the-apple-watch/ (May 2015).

IBM (2017). Connected cars with IBM Watson IoT. https://www.ibm.com/internet-of-things/iotsolutions/iot-automotive/connected-cars/ (accessed on 03 October 2017).

IEEE (2015). Toward a definition of Internet of Things (IoT).

Iliopoulos, M. and Terzopoulos, N. (2016). Wearable miniaturization: dialog's DA14580 bluetooth® smart controller and bosch sensors. Dialog Semiconductor, 7 March 2016.

Khaleel, H.R. (2014). *Innovation in Wearable and Flexible Antennas (book)*. Southampton, UK: WIT Press.

Intel (2016). Data is the new oil in the future of automated driving. *Intel Newsroom* (15 November 2016).

Iqbal, M.H., Aydin, A., Brunckhorst, O. et al. (2016). A review of wearable technology in medicine. *Journal of the Royal Society of Medicine* 109 (10): 372–380.

ISO (2018). ISO/IEC JTC 1/SC 41 - Internet of things and related technologies.

ITU (2012). Overview of the internet of things. https://www.itu.int/rec/T-REC-Y.2060-201206-I. (accessed January 2020).

Jagan Mohan Reddy, N. and Venkareshwarlu, G. (2013). Wireless electronic display board using GSM technology. *International Journal of Electrical, Electronics and Data Communication*, ISSN: 2320-2084 1 (10): 50–54.

Kotz, D., Gunter, C.A., Kumar, S., and Weiner, J.P. (2016). Privacy and security in mobile health – a research agenda. *IEEE Computer* 49 (6): 22–30.

Lymberis, A. and Dittmar, A. (2007). Advanced wearable health systems and applications. *IEEE Engineering in Medicine and Biology Magazine* 26 (3): 29.

Lyons, K. (2015). What can a dumb watch teach a smartwatch? Informing the design of smartwatches. *Proceedings of the 2015 ACM International Symposium on Wearable Computers*, Osaka, Japan, pp. 3–10, 2015. 6.

Mercer, K., Giangregorio, L., Schneider, E. et al. (2016). Acceptance of commercially available wearable activity trackers among adults aged over 50 and with chronic illness: a mixed methods evaluation. *JMIR mHealth and uHealth* 4: e7.

Mitzner, T.L., Boron, J.B., Fausset, C.B. et al. (2010). Older adults talk technology: technology usage and attitudes. *Computers in Human Behavior* 26 (6): 1710–1721.

Motti, V.G. and Caine, K. (2014). Human factors considerations in the design of wearable devices. *Proceedings of the Human Factors and Ergonomics Society Annual Meeting*, Vol. 58. Los Angeles, CA: SAGE Publications, pp. 1820–1824.

NYT (2017). BMW and Volkswagen try to beat Apple and Google at their own game. *The New York Times* (22 June 2017). https://www.nytimes.com/2017/06/22/automobiles/wheels/driverless-cars-big-datavolkswagen-bmw.html.

NYT (2017). The race for self-driving cars. *The New York Times* (6 June 2017). https://www.nytimes.com/interactive/2016/12/14/technology/how-self-driving-carswork.html.

OECD (2012). Machine-to-machine communications: connecting billions of devices. http://www.oecd.org/officialdocuments/publicdisplaydocumentpdf/?cote=DSTI/ICCP/CISP(2011)4/FINAL&docLanguage=En (accessed January 2020).

Pantelopoulos, A. and Bourbakis, N.G. (2010). A survey on wearable sensor-based systems for health monitoring and prognosis. *IEEE Transactions on Systems, Man, and Cybernetics, Part C (Applications and Reviews)* 40 (1): 1–12.

Popat, K.A. and Dr. Sharma, P. (2013). Wearable computer applications a future perspective. *International Journal of Engineering and Innovative Technology (IJEIT)* 3 (1).

Ronen, E., Shamir, A., Weingarten, A.-O., and O'Flynn, C. (2017). IoT goes nuclear: creating a ZigBee chain reaction. *IEEE Security and Privacy (S&P)*, San Jose, CA, USA.

Santa Detector (2018). IFTTT. https://ifttt.com/applets/170037p-santa-detector (accessed January 2020).

Schwartz, E.J., Avgerinos, T., and Brumley, D. (2010). All you ever wanted to know about dynamic taint analysis and forward symbolic execution (but might have been afraid to ask). *IEEE Security and privacy (S&P)*, Claremont Resort, Berkeley, CA, USA.

Sharir, M. and Pnueli, A. (1981). *Two Approaches to Inter-Procedural Dataflow Analysis*. New York: Computer Science Department, New York University.

Sivaraman, V., Gharakheili, H.H., Vishwanath, A. et al. (2015). Networklevel security and privacy control for smart-home IoT devices. *Wireless and Mobile Computing, Networking and Communications (WiMob)*, Anaheim, California, USA.

SmartThings (2018a). Samsung SmartThings add a little smartness to your things. https://www.smartthings.com/ (accessed March 2020).

SmartThings (2018b). SmartThings community forum for third-party apps. https://community.smartthings.com/ (accessed March 2020).

Stankovic, J. (2014). Research directions for the internet of things. *Internet of Things Journal, IEEE* 1 (1): 3–9.

Starner, T. and Martin, T. (2015). Wearable computing: the new dress code. *Computer* 48 (6): 12–15. 5.

ThingsWorx (2018). PTC: industrial IoT. https://www.ptc.com/en/about (accessed 20 June 2018).

Tian, Y., Zhang, N., Lin, Y.-H. et al. (2017). SmartAuth: user-centered authorization for the internet of things. *USENIX Security Symposium*, Vancouver, BC, Canada.

TREND MICRO (2014). Understanding the Internet of Things (IoT).

UC Website (2015). UC workshop: using Google Glass in class. http://www.canberra.edu.au/aboutuc/media/monitor/2014/may/9-google-glass (accessed 25 June 2015).

United States Department of Defense (2016). DoD policy recommendations for the Internet of Things (IoT). https://www.hsdl.org/?abstract&did=799676 (accessed 09 April 2018).

United States Government Accountability Office (2017). Internet of things status and implications of an increasingly connected world. https://www.gao.gov/assets/690/684590.pdf (accessed December 2019).

Vallée-Rai, R., Co, P., Gagnon, E. et al. (1999). Soot: a Java bytecode optimization framework. *Centre for Advanced Studies on Collaborative Research*, Mississauga, Ontario, Canada.

WEF (2015). Industrial internet of things: unleashing the potential of connected products and services. http://www3.weforum.org/docs/WEFUSA_IndustrialInternet_Report2015.pdf (accessed March 2020).

Wright, A. (2017). Mapping the internet of things.

Wright, R. and Keith, L. (2014). Wearable technology: if the tech fits, wear it. *Journal of Electronic Resources in Medical Libraries* 11 (4): 204–216. 3.

Yashiro, T. (2013). An Internet of Things (IoT) architecture for embedded appliances. *Humanitarian Technology Conference (R10-HTC), 2013 IEEE Region 10*. Sendai, Miyagi, Japan: IEEE.

3

Architectures

3.1 Introduction

Complexity is one of the biggest challenges that face the designer when planning an IoT or wearable solution. A characteristic solution involves a number of heterogeneous IoT devices, with sensors that generate data which is then analyzed to provide insights. Further, a myriad of IoT and wearable devices are connected through a gateway device to a network. The job of a gateway is to enable the devices to communicate with each other and with Cloud services and applications. Thus, we need to develop a process flow for a concrete framework over which an IoT or a wearable solution is built.

The architecture portrays the structure of IoT and wearable solutions including the physical aspects (i.e. devices, sensors, actuators) and the virtual aspects (i.e. services, protocols).

There is no single IoT architecture that is agreed upon universally by the technical communities. Various architectures have been proposed by different researchers and technical bodies. However, adopting a multilayered architecture allows the designer to focus on improving the understanding about how all of the aspects of the architecture operate independently before they are integrated into an application. Such modular approach supports managing the complexity of the IoT and wearable solutions.

For data-driven IoT applications, a basic three-tiered architecture, which will be discussed later in this Chapter, can be used to understand the flow of information from smart devices, through a networking element(s), and out to the Cloud services. A more elaborate IoT architecture would include additional vertical layers that cut across the other layers, such as data management and information security.

Fundamentals of IoT and Wearable Technology Design, First Edition. Haider Raad.

Published 2021 by John Wiley & Sons, Inc.

In this Chapter, various architectures used in IoT and wearable devices along with important architecture concepts will be discussed. Further, simplified and versatile architectures are proposed to help the reader articulate the key functions and elements of IoT and wearable devices.

3.2 IoT and Wearable Technology Architectures

3.2.1 Introduction

Because of the outstanding opportunities IoT and wearable devices promise, more enterprises call for their inclusion in their business and processes. However, no proliferating technology has ever grown without adhering to certain standards. Hence, establishing reliable architectures for IoT and wearable technology becomes inevitable.

IoT and wearable technology protocols and platforms are in a state of flux; how these technologies grow and what options emerge for innovative designs will have a tremendous effect on how they expand as we move forward.

As most designers know, even the simplest project requires careful planning and an architecture that comply with a set of standards. Furthermore, when projects become more complex, detailed architectural plans are often required by law.

IT network architectures have evolved significantly over the past 15 years and are generally well-developed and understood; however, the network architectures of IoT and wearable technology are new and need a fresh perspective. It is worth noting that while some similarities between the network architectures of IT and connected devices do exist, in most cases, the challenges and requirements of IoT and wearable systems greatly differ from those of conventional IT networks.

IT networks are essentially concerned with the infrastructure that transports data, regardless of its type. The main goal of IT networks is the reliable and uninterrupted support of enterprise applications such as email, websites, and databases. On the other hand, networks of connected devices are about the data generated by sensors and how it is used. Thus, the core of such architectures is about how the data is transported, aggregated, processed, and eventually acted upon.

3.2.1.1 The Motivations Behind New Architectures

Typically, the scale of IT network is in the order of a few thousand devices (i.e. printers, laptops, servers, hand-held computers). The conventional networking model, with architectures for wide area network (WAN), Wi-Fi, data centers, and so on, is well defined. But when the scale of a network goes from a few thousand nodes to several millions, then this is a different story. In many scenarios, IoT would introduce a model where a utility such as in IIoT, IoV, and smart cities could easily be required to support a network of such scale. Obviously, Internet Protocol version 6 (IPv6) is the natural foundation for networks with a scale of this order.

Moreover, conventional models of IT security are certainly not suitable for the new attacks connected devices (i.e. IoT, wearables) are prone to. Connected devices require rigorous mechanisms of authentication, encryption, and intrusion prevention that match the dynamics of industrial protocols and are capable of responding to attacks on critical infrastructure. However, the endpoints of connected devices are usually located in wireless sensor networks that are operated by unlicensed bands and are visible to the world through spectrum analysis equipment.

There will be a massive amount of data generated by connected devices. Although most data generated by such devices is unstructured, the insights it provides through analytics can radically transform processes and is able to create new business models. However, such vast data could become difficult to be accommodated and analyzed effectively. Hence, unlike IT networks, connected devices should be designed to handle data consumption throughout the architecture itself by filtering and reducing unnecessary data traveling upstream to provide the fastest possible response.

Lastly, most sensors in connected devices are designed to perform a single task, and they are typically small, inexpensive with limited power, processing, and memory resources. They often have a low duty cycle meaning that their transmission duration is small compared to their "idle" time, unless when there is a major event to report.

Because of the substantial scale of these devices and the large heterogeneous environments where they are typically deployed, the networks that provide connectivity tend to support very low data rates compared to IT networks, which enjoy connection speeds in the orders of gigabits per second (Gbps) and nodes with powerful CPUs. Thus, connected devices require a new class of connectivity technologies that meet both the scale and power constraint limitations.

3.2.1.1.1 Centralized vs. Decentralized Network At the beginning of the IoT journey, the architecture was not adequately equipped for the surge of data transfer that will accompany the exponential growth of connected devices. The centralized layout of data handling was not sufficiently robust or fast for real-time processing needs of IoT and wearable systems.

3.2.1.1.2 What Is the Difference Between Centralized and Decentralized Networks? From the consumer point of view, the Internet is based on centralized servers which receive, process, and return data to the end user. Centralized systems are now overwhelmed with data processing needs with the exponential rise in real-time data processing required by connected devices. A decentralized network is needed, and a decentralized Internet is crucial to the evolution of connected devices.

In general, a decentralized network architecture distributes workloads among several entities, instead of relying on a single entity such as a central server. This trend is enabled thanks to the rapid improvements in the computational power of microprocessors which now offer a performance well beyond the needs of most applications of connected devices.

Analysts at Gartner, Inc. stated that 8.4 billion devices were connected to the Internet in 2017, a rise of 31% from the previous year. They also expected this number to reach 20.4 billion by 2020. The number is expected to balloon further to 500 billion by 2030, according to a market analysis by Cisco.

A self-driving car generates roughly ten gigabytes of data per mile. If self-driving vehicles continue to grow in number, it will be impossible to send data to centralized servers for processing every time a vehicle encounters a stop sign or a pedestrian. A microsecond of time is of significant importance in such scenarios. Here is where Edge computing comes into play.

3.2.1.2 Edge Computing

The increasing number of IoT devices at the edge of the network (i.e. closer to the source of the data being generated) is giving rise to an enormous amount of data to be computed at the data centers, pushing network bandwidth requirements to the limit.

The aim of Edge Computing is to bring computing, and data filtering and storage closer to the devices where it's being collected, rather than relying on a central site that can be thousands of miles away. This is done so that data do not suffer from latency issues that can affect an application's performance. Moreover, enterprises can save money by having the processing performed locally, reducing the amount of data that needs to be processed at the Cloud.

Edge Computing describes the work that happens at the edges of the IoT network, where the physical devices exploit mobile phones, smart devices, and/or network gateways to perform tasks and provide services on behalf of the Cloud. With an emphasis on reducing latency, improving privacy and security, and minimizing bandwidth costs within data-driven IoT applications, Edge computing architectures are becoming increasingly common in the realm of IoT and wearable devices.

Fog and Mist computing, which are relatively new terms specific to IoT, also refer to extending computing to the edge of the network. Let's find out the main differences between these terms.

3.2.1.3 Cloud, Fog, and Mist

3.2.1.3.1 Cloud Computing In IT networks, the data used by a node or a server is typically generated by the client/server communications model, and it satisfies the needs of the application. In sensor networks, however, most of the generated data lacks structure and is of little use on its own.

A logical space for such activity is the Cloud where data processing is centralized. The main advantage of this model is simplicity. Smart things only need to connect to a central cloud application where all IoT nodes are overseen and all analytics are processed. However, as data volume along with the number of heterogeneous objects connecting to the network increase, new requirements emerge such as minimizing processing latency, conserving bandwidth traffic, and increasing efficiency. As mentioned previously, these requirements drive the need for data analysis much closer to the IoT node.

Thus, an important consideration is to think of an IoT architecture design that is capable of handling this amount of data in an efficient way such that it can be swiftly analyzed and lead to the envisioned business benefits. The volume of data generated by IoT nodes can be large enough that it can inundate the Cloud. For example, a network of 1 million smart meters will generate close to 1 TB of data per day. This could easily pose a challenge for the network and application server that are not prepared to deal with this amount of data traffic, analysis, and storage.

3.2.1.3.2 Fog Computing The solution to the abovementioned challenges is to decentralize data management and distribute it throughout the IoT system, between the Cloud and endpoint. Within the realm of IoT, this decentralization is widely known as Fog computing. A Fog node can be any device with computing and storage capability, and network connectivity. Switches, routers, gateways, and servers are some examples of Fog nodes. Instead of making the journey to the Cloud, analyzing IoT data close to where it is generated could minimize latency, and offloads substantial traffic from the network.

An advantage of this redistribution is that the Fog node allows control and analytics closer to the endpoint which gives rise to better performance over constrained networks. Another advantage of this is that the Fog node has circumstantial awareness of the sensors it is handling due to its proximity. For example, there might be a Fog router on a hydroponic garden that monitors all the sensor activity in that garden. Because the Fog node is capable of analyzing information from all the sensors in that garden, it can provide contextual analytics of the data it is receiving and may decide to dispatch only the relevant/important information to the Cloud over the backhaul network. Thus, a great reduction of data volume sent to the Cloud is achieved.[1]

3.2.1.3.3 Mist Computing As mentioned in the previous section, a natural place for a Fog node is in the network device that sits within a close proximity to the IoT device, and these nodes are typically spread throughout the network. More recently, however, the concept of Fog computing has been pushed even further to the extreme edge of IoT network and in many cases sits directly in the sensors and IoT endpoints themselves. As mentioned previously, if a Cloud exists in the sky, and Fog resides near the ground, then Mist is what actually sits on the ground.

Fog, Mist, and Edge computing could be viewed as interchangeable terms in the literature, but we would rather refer to Edge computing as a concept whereas Fog and Mist as standards. One could think of Fog as the upper part of the Edge, while Mist represents the lower part. Thus, the concept of Mist is to push computing to the furthest point possible, right into the IoT device or sensor itself. Thanks to rapid advancements in microprocessors and sensor technologies, some new generations of IoT endpoints have sufficient capabilities to perform basic computations, filtering, and low-level analytics to make initial decisions. For example, consider a smart city setting where level sensors are used in public trash cans. While a Fog node residing on an electrical pole at the distribution network may have an excellent view of all the trash cans on a certain avenue, a node on each can would have clear view of a garbage level and would be able to swiftly issue an alert. The Fog node, on the other hand, would have a broader perspective and would be able to decide whether a garbage truck is needed because a large number of cans are filled due to, for instance, a heavy pedestrian traffic. Lastly, a more refined data is sent to the Cloud for historical and statistical analysis.

It is important, however, to emphasize that Fog and Mist computing in no way could take the place of the Cloud. They quite complement each other, and a plethora of applications actually require firm cooperation between these layers. Mist and Fog layers simply act as a first resort for filtering, and analysis which saves the Cloud from being accessed in each and every event. Thus, only time-sensitive data are analyzed on the Mist or Fog node closest to the endpoints generating the data. Data that can wait longer for analysis (i.e. seconds or minutes) is forwarded to an aggregation node and then sent to the Cloud for big data analytics and long-term

---

1 The concept of Fog computing was first coined by Flavio Bonomi and Rodolfo Milito of Cisco Systems. The term gets its name from a comparative analogy to Cloud computing. If the Cloud exists at a higher layer (in the sky), then the Fog layer resides closer to the physical layer (near the ground). An interesting fact is that the term "Fog" was actually suggested by Ginny Nichols, Milito's wife. Although she was not involved in this project, she had a great understanding of what the team was developing. One day she said: "why don't we call it the "fog" layer? And, there it was!

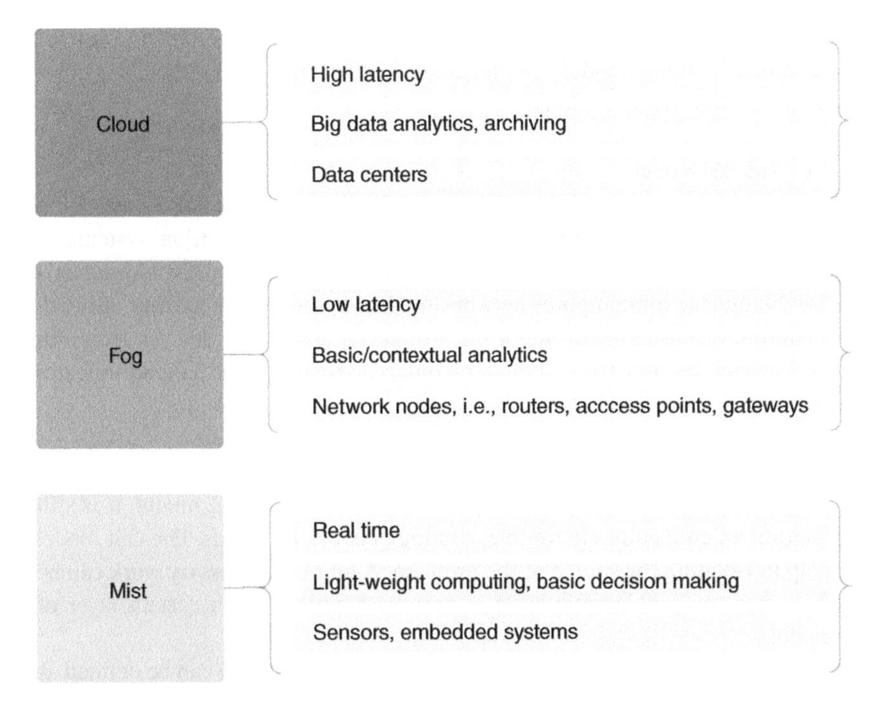

Figure 3.1 A comparison between Cloud, Fog, and Mist computing.

storage. To conclude, when thinking of designing an IoT network, one should not only consider the amount of data to be analyzed, but also the time sensitivity of this data. Understanding these considerations will help in deciding whether Cloud computing is enough or whether Mist or Fog computing would improve the efficiency of the system. In essence, Fog and Mist are standards that enable reproducible structure in the Edge computing concept, so businesses can push the computing out of Clouds for a more efficient and more scalable performance. Figure 3.1 summarizes the differences between Cloud, Fog, and Mist.

3.2.2 IoT Architectures

The new challenges and requirements of IoT are driving an entirely new area of network architecture. In the past decade, architectural standards and frameworks have materialized to address the challenge of deploying extremely large-scale IoT networks. The underlying concept in all these architectures is to support data, processes, and functions that IoT devices would perform. In the following sections, some of the widely known architectures will be discussed.

However, the reader must first be familiar with the Open Systems Interconnection (OSI) model in order to understand the IoT architectures presented in the following sections.

3.2.2.1 The OSI Model

The OSI model is a conceptual model created by the International Organization for Standardization which enables heterogeneous communication systems to communicate using standardized protocols. The OSI model can be viewed as a universal language for computer networking. It is based on the concept of dividing a communication system into a stack of seven abstract layers. As shown in Figure 3.2, each layer of the OSI model handles a specific job and communicates with the layers above and below.

3.2.2.2 Why Does the OSI Model Matter?

Although the modern Internet does not strictly follow the OSI model, it is still very helpful when it comes to troubleshooting network problems. The OSI model can help in isolating the source of the problem. A lot of unnecessary work can be avoided once the problematic issue is narrowed down to one certain layer of the model.

The seven abstraction layers of the OSI model (top to bottom) can be defined as follows:

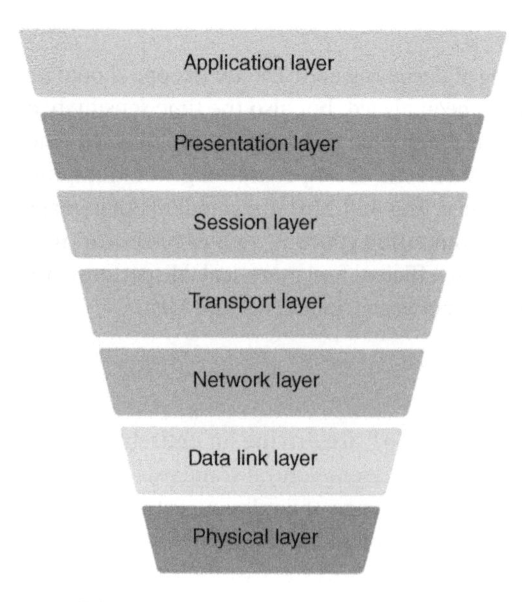

Figure 3.2 The OSI reference model.

3.2.2.2.1 7. The Application Layer This is the only layer that interacts with the user's data directly. Software applications rely on this layer to actuate communications. However, it should be noted that client software applications (i.e. web browser, email clients) are not part of the application layer; rather the application layer is responsible for the governing protocols and manipulating the data that the software application relies on to present meaningful data to the user. Application layer protocols include Hypertext Transfer Protocol (HTTP) as well as Simple Mail Transfer Protocol (SMTP).

3.2.2.2.2 6. The Presentation Layer This layer is responsible for data conditioning including data translation, encryption, and compression, in order to be used by the application layer. For example, if two communicating devices are using different encoding methods, this layer takes care of translating the incoming data into a syntax comprehensible by the application layer of the receiver. Another example is that if the devices are communicating over an encrypted connection, the presentation layer would be responsible for integrating the encryption on the sender's end in addition to decoding the encryption on the receiver's end so that a readable data can be presented at the application layer.

This layer is also responsible for data compression when it is received from the application layer before forwarding it to the layer below (session layer), which promotes communication speed and efficiency.

3.2.2.2.3 5. The Session Layer This layer is primarily responsible for enabling and terminating communication between the two devices. The time between when the communication is enabled and terminated is known as "session." The session layer ensures that the session stays enabled as long as the data is being exchanged; this layer is also responsible for data transfer synchronization.

3.2.2.2.4 4. The Transport Layer This layer is responsible for an end-to-end communication over a network. This involves taking data from the session layer and breaking it up into smaller segments before delivering it to the layer below (network layer). On the receiving end, the transport layer is responsible for re-assembling these segments into data the session layer can utilize. The transport layer also performs flow and error control by ensuring that the data received is complete, and requesting a re-transmission if this is not the case.

3.2.2.2.5 3. The Network Layer The network layer is in charge of facilitating data transfer between two separate networks. This layer is unnecessary if the devices communicating are on the same network. The network layer breaks up the segments received from the transport layer into smaller chunks, called packets, and reassembles these packets on the receiving device. The network layer is also in charge of routing, i.e., finding the best physical path for the data for delivery to its final destination.

3.2.2.2.6 2. The Data Link Layer Data link handles the flow of data into and out of a physical link in a network identifying and correcting errors occurred in the physical layer. As mentioned previously, the network layer is unnecessary when the two communicating devices are on the same network. The data link layer comes into play when the two devices are on the same network. This layer takes packets from the network layer and breaks them into smaller fragments called frames. Flow and error control are also performed in this layer.

3.2.2.2.7 1. The Physical Layer The physical layer includes the actual hardware used in the data transfer, such as the cables, switches, and network interface cards (NICs). This is also where the data gets converted into a stream of binary bits.

3.2.2.3 Data Flow Across the OSI Model

For human-readable information to be transported from one device to another over a network, the data must travel down the seven OSI layers on the sending device and then travel up the seven layers on the receiving one.

For example: User A wants to send User B an email. User A composes a message in an email application client on their laptop and then presses the "send" button. The email application (i.e. Microsoft Outlook) will pass the message over to the application layer, which will use SMTP and forward the data to the presentation layer. The presentation layer will in turn compress the data and forward it to the session layer, which will initiate the communication session.

The data are then segmented at the sender's transport layer. These segments will be broken up into packets at the network layer where the optimal path is determined. The segments are then broken down even further into frames at the data link layer. The data link layer then delivers those frames to the physical layer, where data are converted into a stream of binary bits to be transferred through a physical medium such as a cable or an antenna.

Once User B's computer receives the bit stream through the physical equipment, the data will flow through the same series of layers on their end, but in the opposite order, all the way to the application layer which will feed the human-readable data along to User B's email software application, allowing User **A**'s email to be read on User B's laptop screen.

3.2.2.4 Common IoT Architectures

While we cannot cover all of the architectures reported in the literature, the following list should give the reader a solid understanding of the core design considerations and typical layers in an end-to-end IoT stack.

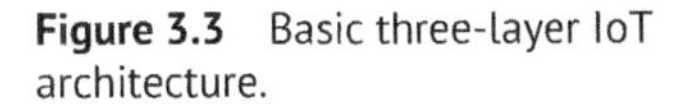
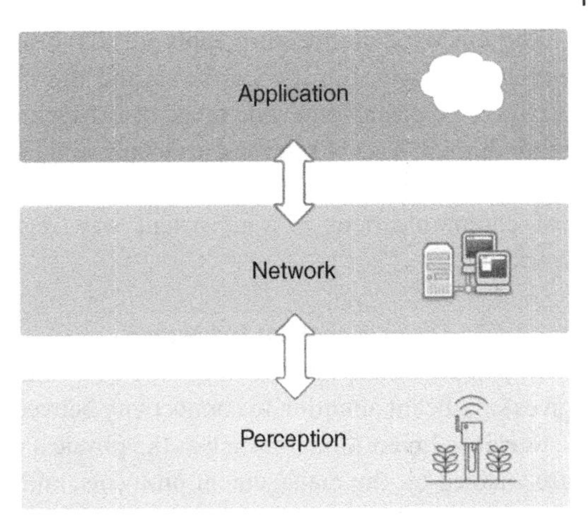

Figure 3.3 Basic three-layer IoT architecture.

*3.2.2.4.1 **Basic Three-Layer IoT Architecture*** While there is a multitude of components that build a complete end-to-end IoT architecture, this one simplifies it down to three fundamental blocks:

Perception layer: Sensors, actuators, and embedded systems that interact with the environment.

Network Layer: This is where data are aggregated, devices are connected over a network, and information is routed to the application layer.

Application Layer: Data processing, analytics, and archiving.

Figure 3.3 depicts an overview of the basic three-tier IoT architecture.

*3.2.2.4.2 **oneM2M Architecture*** The aim of the M2M Technical Committee which was created by the European Telecommunications Standards Institute (ETSI) in 2008 was to establish a common architecture that would lead to an accelerated adoption of M2M devices. A few years later, the scope has expanded to include IoT. Other organizations also started to create their own versions of architectures, which triggered the need for a unified standard for M2M. Recognizing this need, ETSI and other founding entities launched oneM2M architecture in 2012 as a global initiative designed to advance efficient M2M and IoT communication systems. The main goal of this architecture is to establish a common services layer, which can be embedded in field devices to enable communication with the application servers. This architecture emphasizes on the services, platforms, and applications of IoT which include telemedicine, IoV, smart grids, and smart city automation.

By developing a horizontal platform architecture, oneM2M promotes standards that solve one of the main challenges in designing an IoT architecture: Interoperability, i.e. dealing with the diversity of devices, operating systems, and access methods.

The oneM2M architecture splits up IoT functions into three key layers: the application layer, the services layer, and the network layer. While simple, this architecture promotes a wide range of IoT technologies and supports interoperability through a set of routines, protocols, and tools for building software applications. Such framework and Application Program Interface (API) allow end-to-end IoT communications in a consistent way, regardless of how diverse the networks are.

3.2.2.4.3 Layers of oneM2M Architecture

Applications layer: This layer includes the application layer protocols and gives significant attention to connectivity between devices and their applications.

Services layer: Modules such as the physical network that the IoT applications are enabled by, the management protocols, and all hardware are represented by this layer.

The physical portion of the communications network between the central backbone and the individual local networks such as the cellular, Multiprotocol Label Switching (MPLS)[2] networks, virtual private networks (VPNs) are a few examples. An interfacing sub-layer also exists in this layer which adds APIs and middleware that enable third-party services and applications. One of the goals of this architecture is to "develop technical specifications which address the need for a common M2M Service Layer that can be readily embedded within various hardware and software nodes, and rely upon connecting the myriad of devices in the field area network to M2M application servers, which typically reside in a Cloud or data center."

Network layer: This layer includes the communication devices and networks that link them. Manifestations of such communications infrastructure include wireless mesh technologies, wireless point-to-multipoint systems, and wired device connections.

3.2.2.4.4 The IoT World Forum (IoTWF) Architecture In 2014, a seven-layer IoT architectural reference model was published by the IoTWF architectural committee which was led by Cisco, IBM, Rockwell Automation, and other key players in the industry. This architecture offers a polished, yet simplified perspective on IoT. More importantly, it includes Edge computing, data storage, and accessibility. It also offers a concise way of visualizing IoT from a technical point of view. Each of the seven layers is subdivided into specific functions, and

---

2 MPLS is a routing technique in telecommunications networks that transports data from one node to another based on short path labels instead of network addresses, thus avoiding complex lookups in the routing table and increasing the speed of traffic flows.

security is encompassed across the entire model. The IoTWF reference model is shown in Figure 3.4.

The seven layers of the IoTWF reference model are defined as follows:

Layer 1: Physical Devices and Controllers

This layer comprises the "things" in the Internet of Things, which includes the various endpoint devices, sensors, and actuators. The size of these "things" can range from microelectromechanical sensors (MEMS) to massive machines and equipment. Their primary function is generating meaningful data about a process and capability of being controlled over a network.

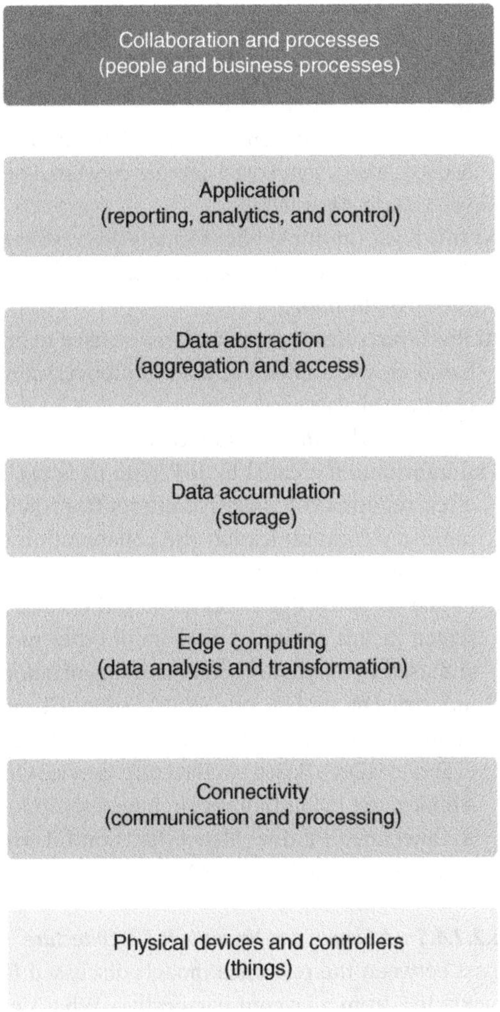

Figure 3.4 IoT reference model reported by the IoT World Forum.

Collaboration and processes
(people and business processes)

Application
(reporting, analytics, and control)

Data abstraction
(aggregation and access)

Data accumulation
(storage)

Edge computing
(data analysis and transformation)

Connectivity
(communication and processing)

Physical devices and controllers
(things)

Layer 2: Connectivity

The primary function of this IoT layer is the reliable and prompt transmission of data. The connectivity layer encompasses all networking elements of IoT, and its functions include communication among Layer 1 entities, switching and routing, protocol translation, and network-level security.

Layer 3: Edge Computing

The focus of this layer is on data reduction and converting data flows in the network into information that is ready for processing and/or storage by higher layers. The characteristic principle of this layer is that the processing of information is initiated as early and as close to the edge of the network as possible. Another important function that takes place at this layer is the assessment of data to decide if it will be filtered or aggregated before forwarding to a higher layer.

Layer 4: Data Accumulation

At this layer, data are accumulated and stored so it could be used by applications when needed. Also, event-based data are converted here to query-based processing.

Layer 5: Data Abstraction

At this layer, multiple data formats are restored, and consistency is ensured for data coming from different sources.

Layer 6: Application

At this layer, software applications are used to interpret data. Reports are provided based on the analysis of data. Moreover, control and monitoring actions take place at this layer.

Layer 7: Collaboration and Processes

The information created by IoT systems is useless unless it yields action, which often requires collaborative efforts from people and processes. At this layer, multistep communication and collaboration occur where application information is shared and consumed.

It should be noted that in addition to the three IoT reference models already presented in this section, a plethora of other models exist. These architectures are endorsed by standards bodies and organizations and are often specific to certain industries or applications. Widely used IoT architectures the reader needs to be aware of include Software Defined Networking (SDN) Based Architecture, Quality of Service (QoS) Based Architecture, Service Oriented Architecture, Mobility First Architecture, CloudThings Architecture, IoT-A Architecture, S-IoT (Social IoT) Architecture, Purdue Model for Control Hierarchy, and Industrial Internet Reference Architecture (IIRA).

3.2.2.4.5 A Simple and Versatile IoT Architecture Although significant differences exist between the reference models discussed in the previous section, they each tackle IoT from a layered perspective. What's common between these models is that they all acknowledge the interconnection of the "things" in IoT to a network that transports the data that will be eventually used by applications, in the cloud, or somewhere in between.

It is not the intention of this book to promote any of the aforementioned IoT architectures, and it should also be noted that IoT reference models may vary based on the industry, application, or the technology being deployed.

In this section, an IoT reference model that highlights the primary building blocks common to most IoT frameworks is presented, which is intended to help the reader in designing an IoT product.

In essence, IoT reference models include numerous elements such as sensors, actuators, protocols, and services. The framework presented here is a simplified IoT architecture that contains the most basic building blocks (layers) which can be used as a foundation to understand key design and deployment principles. These basic layers can be expanded on as needed based on the industry-specific use cases. If needed, the reader is referred to one of the more complex reference models covered in this book or in the literature.

3.2.2.5 Layer 1: Perception and Actuation (Sensors and Actuators)

This layer includes the sensing or "perception" system where data is collected from the environment, a process, or a smart object performing a certain function, to be used in decision making or analytics. An actuator can also be present at this layer that allows the smart objects to perform an action (i.e. to switch on or off a light, adjust an airflow of a valve, to increase or decrease an engine's rotation speed and more).

Some limited data processing (Mist computing) can occur at this layer if an immediate response is necessary. However, if more in-depth processing of data is needed, which requires more computational power and time then the data needs to be moved to the upper layers (Fog, Cloud/Data centers).

3.2.2.6 Layer 2: Data Conditioning and Linking (Aggregation, Digitization, and Forwarding)

One can think of this layer to have two ends. One end which sits in close proximity to the sensors and actuators, is responsible for aggregating the data and converting it into digital streams (analog to digital conversion) for further processing in the upper layers. The other end, which sits closer to layer 3, contains the Internet gateway which receives the aggregated and digitized data and routes it over a wireless or wired network to the upper layers for further processing.

The significant importance of this stage lies in preprocessing the voluminous amount of analog information collected from the sensors in the previous stage and compressing it to the optimal size for further analysis. Moreover, the aggregated analog data have specific timing and structural characteristics that require special treatment.

Data travel back and forth from smart things to the Cloud through gateways. A gateway is a part of the IoT and wearable technology solution that provides connectivity between the smart things and the Cloud. To reduce the volume of the data aggregated from the sensors, preprocessing and filtering are also performed here before the data is moved to the Cloud. Control commands going from the Cloud to the actuators of the smart things are also passed through the gateway. Using gateways prolongs battery life and lowers latency. Gateways also enable connecting devices without a direct Internet access and provide an additional layer of security by protecting data moving in both directions.

Advanced gateways could have additional capabilities such as analytics, security, and data management services. Although providing insights from the data is somehow less immediate at the gateway than it would be when delivered directly from the sensor-actuator area, the gateway has the computational power to render the information in a more useful form to the end user.

3.2.2.7 Layer 3: Network Transport (Preprocessing, Preliminary Analytics, and Routing)

Once IoT data have been digitized and aggregated, it is ready to travel to the Cloud or data center. However, the data may require further processing before this step is initiated. As infrastructure can be physically located closer to the data source, it is easier and faster to act on the IoT material in or close to real time and provide an output. Thus, only the larger pieces of data that need the computational and analytical power of the Cloud would be forwarded. By minimizing network utilization, security can be significantly enhanced, while improved response times and reduced bandwidth consumption could contribute to more efficient IoT systems. For example, rather than forwarding raw vibration data from pumps, data can be filtered, and preprocessed, then only projection data are sent as to when each pump will fail or need service.

The data at this level may be communicated to a data center, a cloud, or to another IoT device on the network. To allow for such communication diversity, an open and standard-based network protocol with specific characteristics needs to be implemented to accommodate diverse industries and various media structures. Accommodating an enormous number of sensors in a single network along with data security are also required here.

IP is flexible enough to be embedded in smart objects of very different classes, and to exchange data over very different media and network types.

Multiple protocols have been created to optimize IoT data communications. Some networks are based on a push model (a sensor reports to the application at a regular interval or based on a local stimulation), whereas other networks rely on a pull model (where an application queries the sensor); other hybrid approaches are also possible. One very common protocol used in IoT products is Message Queue Telemetry

Transport (MQTT), which is a standard lightweight, publish–subscribe network protocol that transports messages between devices. The sensor can be set up to be a publisher (publishes the required information), the application can be set as the subscriber (receives the information), and an intermediate system is set up to act as a broker to relay the information between the publisher and the subscriber.

3.2.2.8 Layer 4: Application (Analytics, Control, and Archiving)

Non time-sensitive data that needs more in-depth processing, gets forwarded to this layer (typically, a data center or Cloud-based system), to be analyzed, managed, and securely archived. This layer is also used to control the smart objects when necessary. Moreover, this layer instructs smart objects to adapt to certain conditions based on deeper insights gained from historical analysis. Figure 3.5 illustrates the layers of the simplified IoT reference model and their relevance to the Cloud, Fog, and Mist.

3.2.3 Wearable Device Architecture

One could utilize one of the IoT architectures discussed in the previous section to design sophisticated wearable devices. It should be noted, however, that due to the size constraints, and limited computational power of wearables, a typical system architecture would include a gateway to the Internet (i.e. a smartphone) with a dedicated application used for configuring the device and for processing the perceived data. The gateway device is also used for data visualization purposes where relevant processed data are translated into a graphical representation. The perceived data are forwarded wirelessly using a low power transceiver (i.e. Bluetooth Low Energy (BLE)) for basic processing. A more thorough and insightful data

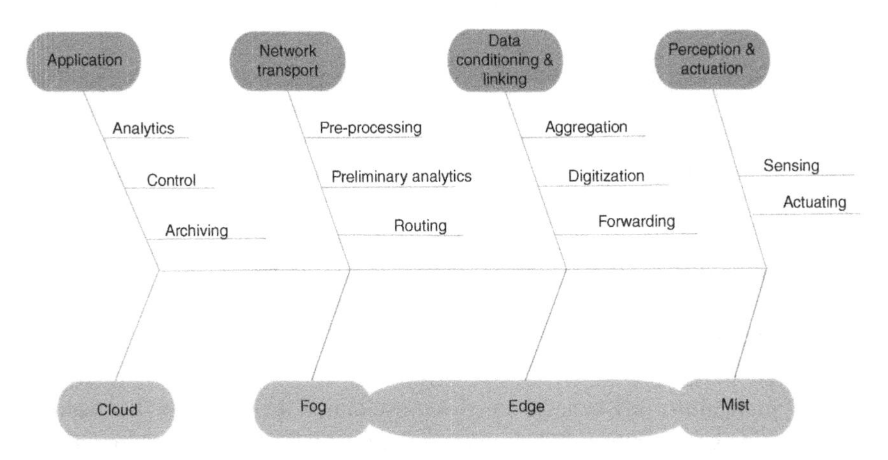

Figure 3.5 Simplified IoT architecture.

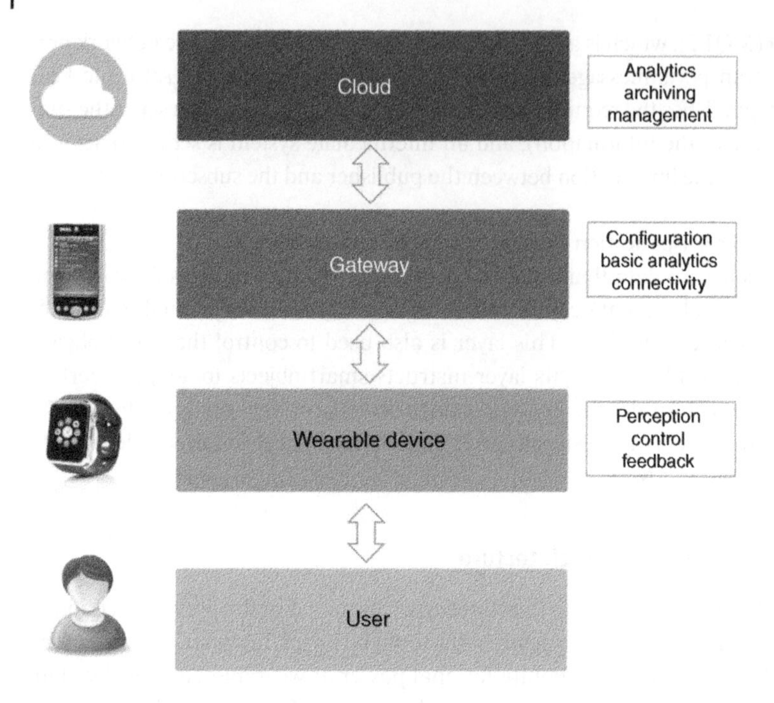

Figure 3.6 A basic architecture for wearable devices.

processing takes place in the Cloud. Processed data are then transferred back to the wearable device for feedback while a copy of it is archived at the data center. Obviously, the gateway provides the means of network connectivity for such back and forth communication. An architecture for a typical wearable device is depicted in Figure 3.6.

It should be noted that wearables can be either simpler or more complex than what's shown in Figure 3.6. For instance, some wearable navigators used by professional hikers connect directly to a Global Positioning System (GPS), thus bypassing the gateway layer. On the other hand, some early fitness trackers relied only on a smartphone for processing and feedback without the need to use a Cloud service.

3.3 Conclusion

The requirements of IoT systems are giving rise to new architectures that address the emerging constraints and data management aspects of IoT. To address these needs, a number of IoT-specific reference models have been proposed in the literature. The features shared between these models are the interconnection of IoT endpoints, the networks that connect them, and the applications that manage them.

Due to the unique requirements of IoT, centralized data management is no longer as practical as it is in traditional IT networks. It is simply not practical to send data to the Cloud over a backhaul connection for processing. Before it is sent to the Cloud, data in IoT systems are filtered, aggregated, and pre-analyzed in layers close to the edge of the network. This drives new concepts: Fog and Mist computing, where services are delivered very close to the IoT endpoints.

This Chapter presented a simplified IoT framework broken down into its most basic building blocks which can be used as a foundation to understand key design and deployment principles that can be applied to industry-specific manifestations. It was also shown that a wearable device can be designed based on an IoT architecture, depending on how complex the functionality is. A basic model is also reported in this Chapter. It was shown that due to various constraints, a typical wearable device architecture would include a gateway device layer which is used for configuring the device and for processing the gathered data, whereas a more in-depth data processing takes place in the Cloud.

Problems

1 Why an architecture is needed for connected devices (IoT and wearables)?

2 What is the difference between centralized and decentralized networks?

3 List three published IoT architectures and research three more from the literature. Compare the six architectures using a table.

4 Give an example of an IoT device and explain its operation using the simplified IoT architecture reported in this chapter.

5 What is Edge Computing? Give four examples of IoT and wearable devices and explain their operation within the context of Edge.

6 What is the difference between Cloud, Fog, and Mist? Explain using two practical examples (one IoT device, and one wearable device).

7 What are the main differences between IT and IoT networks?

8 How is the OSI model related to IoT and wearable technology architectures?

9 Design a basic wearable fitness tracker using the wearables architecture described in this chapter.

10 Design a basic IoT garden monitor using the simplified architecture described in this chapter.

11 Sketch a smart home system and link each component that you use (software and hardware) to an architecture of your choice.

Technical Interview Questions

1 What are the layers of OSI?

2 What is IoT Contiki?

3 What are the pros and cons of MQTT and HTTP?

4 What AWS services are used for Edge computing?

5 What is Microsoft Azure?

6 What is Edge? Why is it popular in IoT and wearables?

7 What is the difference between Cloud and Fog computing?

8 Tell us what you know about Quality of Service (QoS) Based Architecture.

9 Tell us what you know about S-IoT (Social IoT) Architecture.

10 Tell us what you know about Software-defined networking (SDN)-Based Architecture.

Further Reading

Aazam, M., Hung, P.P., and Huh, E.-N. (2014). Smart gateway based communication for cloud of things. *Proceedings of the 9th IEEE International Conference on Intelligent Sensors, Sensor Networks and Information Processing (IEEE ISSNIP '14)*, (April 2014). Singapore: IEEE.

Ahson, S.A. and Ilyas, M. (2011). *Near Field Communications Handbook (Internet and Communications)*. Boca Raton, FL: CRC Press Taylor and Francis, 23 September). ISBN-10: 1420088149.

Akyildiz, I.F., Su, W., Sankarasubramaniam, Y., and Cayirci, E. (2002). Wireless sensor networks: a survey. *Computer Networks* 38: 393–422.

Asghar, M.H. (2015). RFID and EPC as key technology on Internet of Things (IoT). *International Journal of Computer Science and Technology* 6: 121–123.

Atzori, L., Iera, A., and Morabito, G. (2010). The internet of things: a survey. *Computer Networks* 54 (15): 2787–2805.

Bhabad, M. and Sudhir, B. (2015). Internet of things: architecture, security issues and countermeasures. *International Journal of Computers and Applications* 125 (14): 1–4.

Bilal, M. (2017). "A review of internet of things architecture", technologies and analysis smartphone-based attacks against 3D printers. arXiv preprint arXiv:1708.04560, 1–21.

Bormann, C., Lemay, S., Tschofenig, H. et al. (2018). CoAP (Constrained Application Protocol) over TCP TLS and WebSockets, IETF Internet Draft.

Botterman, M. (2009). For the European commission information society and media directorate general, networked enterprise & RFID unit – D4. *Internet of Things: An Early Reality of the Future Internet, Report of the Internet of Things Workshop*, Prague.

Burhanuddin, A.A.-J.M.M.A. (2017). IoT architecture section I: the issue/challenge. *International Journal of Applied Engineering Research* 12: 11055–11061.

Chen, F., Wang, N., German, R., and Dressler, F. (2008). LR-WPAN for industrial applications. *2008 Fifth Annual Conference on Wireless on Demand Network Systems and Services*, Garmisch-Partenkirchen (23–25 January 2008).

Clayman, S. and Gali, A. (2011). INOX: a managed service platform for interconnected smart objects. *Proceedings of the workshop on Internet of Things and Service Platforms (loTSP'11)*, NY, USA, pp. 1–8.

De Deugd, S., Carroll, R., Kelly, K.E. et al. (2006). SODA: Service-oriented device architecture. *IEEE Pervasive Computing* 5: 94–96.

Deng, H. (2008). Research and implementation of the RFID middleware based on SOA [J]. *Journal of Shaanxi Normal University* 10: 1–7.

Dinh, H.T., Lee, C., Niyato, D., and Wang, P. (2013). A survey of mobile cloud computing: architecture, applications, and approaches. *Wireless Communications and Mobile Computing* 13 (18): 1587–1611.

ETSI (2012). oneM2M. www.etsi.org/about/what-we-do/global-collaboration/onem2m.

Garcia-Morchon, O., Rietman, R., Sharma, S. et al. (2016). A comprehensive and lightweight security architecture to secure the IoT throughout the lifecycle of a device based on HIMMO. In: *Algorithms for Sensor Systems. Lecture Notes Computer Science*, vol. 9536 (eds. S. Gilbert, D. Hughes and B. Krishnamachari), 112–128. Cham, Switzerland: Springer.

Guth, J., Breitenbucher, U., Falkenthal, M. et al. (2016). Comparison of IoT platform architectures: a field study based on a reference architecture. *Cloudification of the Internet of Things (CIoT)*, Paris (23–25 November 2016).

Hanes, D. (2017). *IoT Fundamentals*. London, UK: Pearson Education.

Ho, E., Jacobs, T., Meissner, S. et al. (2013). *ARM Testimonials, in Enabling Things to Talk*, 279–322. Berlin, Heidelberg: Springer.

Hunkeler, U., Truong, H.L., and Stanford-Clark, A. (2008). MQTT-S—a publish/subscribe protocol for wireless sensor networks. *Proceedings of the 3rd IEEE/Create-Net International Conference on Communication System Software and Middleware (COMSWARE '08)*, Bangalore, India (January 2008), pp. 791–798.

Jain, A. and Tanwer, A. (2010). Modified Epc global network architecture of internet of things for high load Rfid systems: free download & streaming: internet archive. *Proceedings of International Conference on Advances in Computer Science* 1 (3): 3–7.

Ji, Z., Ganchev, I., and O'Droma, M. (2013). A generic IoT architecture for smart cities. *2014, 25th IET Irish Signals & Systems Conference 2014 and 2014 China-Ireland International Conference on Information and Communications Technologies (ISSC 2014/CIICT 2014)*, Limerick, 26–27 June 2013, pp. 196–199.

Jules, A. (2006). A research survey: RFID security and privacy issue. *Computer Science* 24: 381–394.

Jyothi, T., Vineetha, C., Vandana, J. et al. (2018). WIFI based agriculture environment monitoring system using android mobile application. *National Conference on Emerging Trends in Information, Management and Engineering Sciences*, West Bengal, India, pp. 1–5.

Kang Lee, F.J. and Lanctot, P. (2017). Internet of things: wireless sensor networks. International Electrotechnical Commission.

Kos, A., Pristov, D., Sedlar, U. et al. (2012). Open and scalable IoT platform and its applications for real time access line monitoring and alarm correlation. *Conference on Internet of Things and Smart Spaces. International Conference on Next Generation Wired/Wireless Networking. Lecture Notes in Computer Science*, Springer, Berlin, pp. 22–38.

Lee, B.M. and Ouyang, J. (2014). Intelligent healthcare service by using collaborations IOT personal health device. *International Journal of BioScience and BioTechnology* 6 (1): 155–164.

Locke, D. (2010). MQ telemetry transport (MQTT) v3. 1 protocol specification, IBM developerWorks Technical Library. http://www.ibm.com/developerworks/webservices/library/ws-mqtt/index.html.

Mohammadi, M., Aledhari, M., and Al-Fuqaha, A. (2015). Internet of things: a survey on enabling technologies, protocols and applications. *IEEE Communications Surveys & Tutorials* 17 (4): 2347–2376.

Mukherjee, M., Adhikary, I., Mondal, S. et al. (2017). A vision of IoT: applications challenges and opportunities with Dehradun perspective. *Advances in Intelligent Systems and Computing* 479 (4): 553–559.

Ngu, Q.Z.S.A.H., Gutierrez, M., Metsis, V., and Nepal, S. (2017). IoT middleware: a survey on issues and enabling technologies. *IEEE Internet of Things Journal* 4: 1–20.

Ning, H. and Liu, H. (2012). Cyber-physical-social based security architecture for future internet of things. *Advances in Internet of Things* 2 (1): 1–7.

Ning, H. and Wang, Z. (2011). Future IoT architecture – like mankind neural system or social organization framework. *IEEE Communications Letters* 15 (4): 461–463.

OASIS (2007). Web services business process execution language version 2.0, Working Draft. http://docs.oasis-open.org/wsbpel/2.0/wsbpelspecificationdraft.pdf.

OASIS.org (2014). MQTT version 3.1.1. (OASIS Standard). http://docs.oasis-open.org/mqtt/mqtt/v3.1.1/os/mqttv3.1.1-os.html.

oneM2M (2014). Technical specification. ftp.onem2m.org/Deliverables/20140801_Candidate%20Release/TS-0002-Requirements-V-2014-08.pdf.

Pereira, P.P., Eliasson, J., Kyusakov, R., and Delsing, J. (2013). Enabling cloud connectivity for mobile internet of things applications. *Proceedings IEEE 7th International Symposium on Service Oriented System Engineering (SOSE)*, Redwood City, CA (25–28 March 2013), pp. 518–526.

Qing Hu, X.H. and Shan, Y. (2009). Based on internet of things and RFID middleware technology research. *Micro Computer Information* 25: 105–185.

Ray, P.P. (2014). Internet of things based physical activity monitoring (PAMIoT): an architectural framework to monitor human physical activity. *IEEE CALCON*, Kolkata, pp. 32–34.

Ray, P.P. (2015). Towards an internet of things based architectural framework for defence. *International Conference on Control, Instrumentation, Communication and Computational Technologies (ICCICCT)*, Kumaracoil (18–19 December 2015), pp. 411–416.

Sebastian, S. and Ray, P.P. (2015). When soccer gets connected to internet. *International Conference on Computing and Communication Systems (I3CS), Shillong*, pp. 84–88.

Shelby, Z., Hartke, K., and Bormann, C. (2014). The Constrained Application Protocol (CoAP). *Tech. Rep., IETF 7959.*

Spiess, P. (2009). SOA-based integration of the internet of things in enterprise services. *Proceedings of IEEE ICWS*, Los Angeles, CA (6–10 July 2009).

Stanford-Clark, A. and Linh Truon, H. (2008). MQTT for sensor networks (MQTT-S) protocol specification, International Business Machines Corporation Version 1.

Stojmenovic, I. (2014). Fog computing: a cloud to the ground support for smart things and machine-to-machine networks*Proceedings of the Australasian Telecommunication Networks and Applications Conference (ATNAC '14)*, Melbourne, Australia (November 2014), pp. 117–122.

Toma, I., Simperl, E., and Hench, G. (2009). A joint roadmap for semantic technologies and the internet of things. *Proceedings of the Third STI Road mapping Workshop*, Crete.

Traub, K., Armenio, F., Barthel, H. et al. (2014). The GS1 EPCglobal architecture framework, 1–72, Version 1.6.

Varghese, B. and Buyya, R. (2018). Next generation cloud computing: new trends and research directions. *Future Generation Computational Systems* 79: 849–861.

Villaverde, B.C., Pesch, D., De Paz Alberola, R. et al. (2012). Constrained application protocol for low power embedded networks: a survey. *Proceedings of the 6th International Conference on Innovative Mobile and Internet Services in Ubiquitous Computing (IMIS '12)*, Palermo, Italy (July 2012), pp. 702–707.

Vucinic, M., Tourancheau, B., Rousseau, F. et al. (2014). OSCAR: object security architecture for the internet of things. *Proceeding of IEEE International Symposium on a World of Wireless, Mobile and Multimedia Networks 2014, WoWMoM 2014* (19 June 2014). Sydney: NSW.

Wang, P., Liu, S., Ye, F., and Chen, X. (2018). A fog-based architecture and programming model for IoT applications in the smart grid. Netw. Internet Archit.

Welbourne, E., Battle, L., Cole, G. et al. (2009). Building the internet of things using RFID: the RFID ecosystem experience. *IEEE Internet Computing* 13 (3): 48–55.

Xia, F. (2009). Wireless sensor technologies and applications. *Sensors* 9 (11): 8824–8830.

Yang, G., Xie, L., Mäntysalo, M. et al. (2014). A health-IoT platform based on the integration of intelligent packaging, unobtrusive bio-sensor, and intelligent medicine box. *IEEE Transactions on Industrial Informatics* 10 (4): 2180–2191.

Yannuzzi, M., Milito, R., Serral-Gracia, R. et al. (2014). Key ingredients in an IoT recipe: fog computing, cloud computing, and more fog computing. *Proceedings of the IEEE 19th International Workshop on Computer Aided Modeling and Design of Communication Links and Networks (CAMAD '14)*, Athens, Greece (December 2014), pp. 325–329.

Yunsong Tan, J.H. (2015). A service-oriented IOT middleware model. *Journal of Computer Science* 4: 115–120.

Zhang, W. and Qu, B. (2013). Security architecture of the internet of things oriented to perceptual layer. *International Journal on Computer, Consumer and Control* 2 (2): 37–45.

Zhao, J.C., Zhang, J.F., Feng, Y., and Guo, J.X. (2010). The study and application of the IOT technology in agriculture. *3rd International Conference on Computer Science and Information Technology*, vol. 2, Chengdu (9–11 July 2010), pp. 462–465.

Zhou, L., Xiong, N., Shu, L. et al. (2010). Context-aware multimedia service in heterogeneous networks. *IEEE Intelligent Systems* 25 (2): 40–47.

4
Hardware

4.1 Introduction

The core functionality of IoT and wearable devices starts with data acquired or an action performed by a device. These devices are called endpoints, and they are the "Things" in Internet of Things. The value of IoT and wearable devices is in the data collected by these endpoints, so it is important to understand how they acquire, process, transmit, and receive data.

The designer should ask: What type of sensor, actuator, or microcontroller should be considered for the application/problem in hand? What energy source should be used? What communication module should be considered?

This Chapter highlights the capabilities, characteristics, and functionality of sensors and actuators with an understanding of their limitations and their role in IoT and wearable systems. Criteria for selecting a microprocessor and communication module will be reviewed next. Additionally, deciding on a suitable energy source with a matching application-specific power management design is discussed. Finally, the reader will gain an understanding on how to bring these foundational elements together to realize a smart device that makes most IoT and wearable use cases possible.

4.2 Hardware Components Inside IoT and Wearable Devices

In spite of the variety and types of IoT and wearable devices, the majority share elemental functionalities that must be implemented in the design process. Most smart endpoints have the following components:

- At least one sensor to perceive an analog quantity (typically physical, chemical, biological, or environmental).

Fundamentals of IoT and Wearable Technology Design, First Edition. Haider Raad.

Published 2021 by John Wiley & Sons, Inc.

- A conditioning circuit that filters, amplifies, and converts the perceived signal into a digital one (analog-to-digital conversion (ADC)).
- A processing unit along with a memory and embedded system that serves as the brain power of these smart devices where all computations and processing take place.
- A connectivity unit to transmit the captured data or receive an action command to and from another layer (a mobile phone, cloud, server, etc.) using, typically, one of the wireless communication technologies such as Wi-Fi and Bluetooth.
- Input and output elements for device user interface which may include a button, gesture pad, microphone, camera for input; an LCD display, LED lights, speaker, or other motor-based actuators for output.
- An energy source and most likely a power management system.

Noticeably, the first generation of smart connected devices has been assembled utilizing smartphones components and technologies. For example, in the beginning of the past decade, the sensors and microprocessors of wrist and head-mounted wearable devices have been drafted off smartphones parts. This would, for many, make perfect sense since using tested components with a proven success in a newly introduced technology would help manufacturers balance features, functionality, and price for an uncertain new market. However, electronics manufacturers today are introducing new components designed specifically for wearables. Power-efficient microprocessors and performance-bound hardware of smaller form factor are being driven by the requirements of the next generation of IoT and wearable devices. Figure 4.1 depicts the anatomy of a generic connected device.

4.2.1 Sensors

A sensor is a device that detects a change or an event in an object or environment and converts it to usually an electrical signal. The sensor then forwards the converted signal to a microprocessor unit for analysis to provide a useful output that can be consumed by intelligent devices or humans.

Analog (continuous, nondigital) sensors are very common in biomedical and healthcare devices. Biometric sensors such as heart rate, blood pressure, and electroencephalogram (EEG) are examples of such sensors. It should be noted that an analog front-end (AFE) unit is needed in analog sensors, which is responsible for amplifying, filtering, and conditioning the signal, and converting it to digital, using an ADC, so that it can be processed by a microprocessor.

There are a number of ways to categorize sensors including the following:

a) **Active or Passive:** Active sensors require an external power source to operate, while passive sensors simply detect energy and operate without the need of a power source.

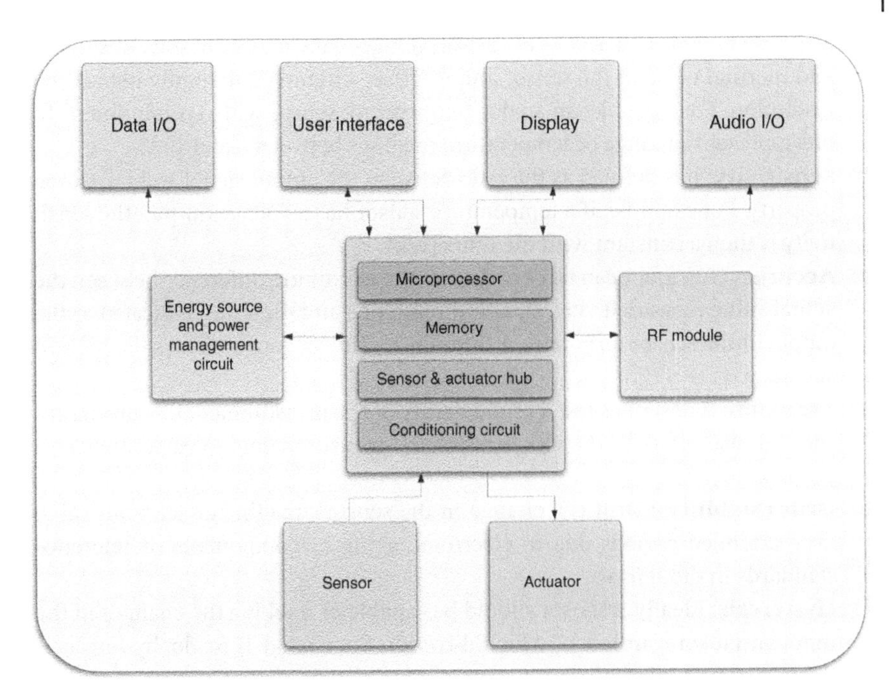

Figure 4.1 Anatomy of a generic smart connected device.

b) **Invasive or Noninvasive:** Invasive sensors are transducers that come into direct contact with the process (i.e. a sensor immersed in a fluid). Noninvasive transducers, on the other hand, do not come into direct contact with the process (i.e. an ultrasonic level sensor).
c) **Deflection or Null:** The signal produces some physical effect (i.e. movement) closely related to the measured quantity such as in a pressure gauge where the value being measured is displayed in terms of the amount of pointer movement. In null type, which is more accurate but more complex than the deflection type, the signal produced by the sensor is counteracted to minimize the deflection. That opposing effect necessary to force a zero deflection has to be proportional to the signal of the quantity to be measured.

Sensors can also be categorized based on their area of application, mechanism, and quantities they measure.

4.2.1.1 Sensor Properties

A sensor should satisfy certain characteristics before integration within a system. Below are typical properties used to characterize sensors:

a) **Resolution:** The resolution of a sensor is defined as the smallest change in the input under test that can be perceived by the sensor. A high-resolution sensor is one that is able to detect a very small change in the input variable. Electronic and thermal noise in the sensor and interface circuitry can highly impact the resolution. For example, an analog temperature sensor with a resolution of 10 bits represents a range of temperature readings between 0 and 1023.

b) **Sensitivity:** It is defined as the ratio between the output signal and measured property. For example, if a temperature sensor has a voltage output, the sensitivity is then a constant with the units [v/k].
c) **Accuracy:** Accuracy can be described as the maximum difference between the actual value measured via a standard reference and the value indicated at the output of the sensor terminals. A difference value of zero indicates the highest accuracy.
d) **Precision:** It describes the reproducibility of the measurement of a given sensor and also refers to the closeness of the measurements to each other in a given scale.
e) **Drift (Stability):** Drift is a change in the sensor's reading or set point value over extended periods due to electronic aging of components or reference standards in the sensor.
f) **Hysteresis:** Ideally, a sensor should be capable of tracking the changes in the input variable regardless of which direction the change is made; hysteresis is the measure of this feature.
g) **Response Time (Responsiveness):** Response time refers to the ability of a sensor to respond to fast changes in inputs.
h) **Dynamic Range:** It refers to the full range from minimum to maximum values a sensor can measure. For example, a given temperature sensor may have a range of −40 to +120 °C.

4.2.1.2 MEMS Sensors

Today, the majority of IoT and wearable devices are based on microelectromechanical sensor (MEMS) technology. MEMS is a technology based on miniaturized electromechanical structures which are made using microfabrication techniques. The dimensions of MEMS devices typically range from one micron to a few millimeters. The dynamics of MEMS are controlled by the microelectronics integrated within the device.

Inertial measurement units (IMUs)[1] found in most modern wristband wearables are perhaps one of the most popular MEMS examples. Depending on the complexity of the wearable device, it could have a single, a few MEMS sensors, or a dedicated sensor fusion interface.

---

1 An **inertial measurement unit (IMU)** is a self-contained electronic device that measures linear and angular motion, force, and magnetic field. It is typically based on a combination of accelerometers, gyroscopes, and magnetometers.

The most challenging aspect in such sensors, however, lies in converting raw data into useful information. As mentioned previously, sensor fusion is the process of collecting multiple output data from multiple sensors to obtain a better total insight. A good example here is the use of data from single 3-axis acceleration along with data from a 6-axis IMU rotation sensor in fitness trackers to have a more accurate information on the user's motion.

4.2.1.3 Commonly Used Sensors in IoT and Wearable Devices

Below is a description of the most common sensors used in IoT and wearable devices:

4.2.1.3.1 Accelerometers An accelerometer is an electromechanical device used to measure static and dynamic acceleration forces due to gravity, motion, and vibration. By measuring acceleration, the angle of the device is oriented with respect to the earth can be found in addition to the direction of motion. In IoT, applications like indoor climate systems and security systems in smart homes are enabled by accelerometers, while in wearables they are the most commonly used sensors. Most modern accelerometers are based on MEMS.

4.2.1.3.2 Gyroscopes Gyroscopes are used to measure and\or maintain rotational motion and angular velocity. They are used to maintain equilibrium and determine direction and orientation of an object.

4.2.1.3.3 Magnetometers A magnetometer is a device that measures magnetic fields. Today's digital compasses are based on magnetometers which provide an orientation relative to the earth's magnetic field. A device equipped with a magnetometer will always have the Magnetic North as a reference.

4.2.1.3.4 Hall Effect Sensors The Hall effect sensor is used to detect a magnetic field. The Hall element is comprised of a thin sheet of an electric conductor with output terminals perpendicular to the direction of current flow. When a magnetic field is present, it produces an output voltage proportional to its strength. The voltage is in microvolts; hence, it needs to be amplified to be of use. Beside a magnetic field, the sensor can be used as a sensor for current, temperature, pressure, etc.

4.2.1.3.5 Altimeters Most altitude sensors can determine altitudes based on the atmospheric pressure. Besides determining the user's altitude, altimeters have higher processing accuracies when implemented in fitness trackers. For example, the altimeter enables fitness trackers to determine whether the user is climbing stairs through sensing the height changes, which allows for a more realistic calorie loss calculation.

4.2.1.3.6 Flex Sensors Flex sensors are passive resistive devices that changes resistance when bent which can be used to detect flexing or bending of an object.

4.2.1.3.7 Galvanic Skin Response (GSR) Sensors Galvanic skin response (GSR) measures the continuous variation of electrical impedance of human skin which enables the detection of psychological, emotional, and physiological parameters.

4.2.1.3.8 Temperature Sensors Low-power temperature sensors drive an electric signal (voltage) that is proportional to the ambient temperature. Most small factor temperature sensors are based on thermocouples, temperature-dependent resistors, often called thermistors, or temperature-dependent transistors.

4.2.1.3.9 Biochemical Sensors Integrating biochemical sensors within IoT and wearable technology is the focus of many companies who are looking for innovative ways to capture and analyze new health-related data.

Research has shown that monitoring the sodium concentration in human sweat serves as an indicator of the person's electrolytic balance and general well-being. To that point, recent studies report the potentials of epidermal transfer tattoo-based potentiometric sensor attached to a compact wireless transmitter for noninvasive sweat monitoring. Another recent study reports a sensor that measures the acetone concentration in breath which is released as a metabolism byproduct to determine whether the user is burning fat.

4.2.1.3.10 Electroencephalograph (EEG) Sensors The EEG sensor is essentially a signal amplifier for detecting a brain's electrical activity from the head's surface where the neurons generate extremely small amplitudes of voltage.

4.2.1.3.11 Optical Heart Rate Sensors Based on the pulse oximetry technique, these optical sensors distinguish between the optical features of the oxygenated and de-oxygenated hemoglobin. The monitor consists of a red LED and optical detector which measures the light reflectance or absorbance during the oxygenation and de-oxygenation cycle, and the heart rate is then determined.

4.2.1.3.12 Gesture Sensors Gesture sensors aim at enhancing the user interface in an electronic device which is usually achieved by enabling a coherent display and touchless communication. Most modern gesture sensors used in wearables utilize four directional photodiodes to sense infrared energy to convert direction, distance, and velocity information to digital information. Other gestures sensors are based on an RGBC sensor, which provides red, green, blue, and clear light sensing which in turn detects light intensity under different lighting conditions. Digital ambient light sensing (DALS), on the other hand, incorporates a photodiode, an amplifier, and an analog-to-digital convertor, in a single chip.

4.2.1.3.13 Proximity Sensors Proximity sensors detect the presence of objects without a physical contact and produce an output in the form of an electromagnetic field or electric signal, while analyzing changes in the return signal. These sensors use light, sound, or ultrasonic sensitive components to detect objects and consist of an emitter and a receiver.

4.2.1.3.14 Capacitive and Inductive Sensors These sensors are based on a high frequency oscillator that creates a field in the immediate proximity of the sensing surface. The presence of an object in this proximity creates a change of the oscillation amplitude where the positive and negative peaks are identified by another unit that triggers the change of the sensor's output. The operation of many sensors and interface components is based on capacitive and inductive sensors, including position, humidity, fluid level, and acceleration sensors, in addition to trackpads and touch screens.

4.2.1.3.15 Passive Infrared (PIR) A passive infrared sensor is used to measure infrared light radiating from objects. They are commonly used in security alarms and auto lighting systems.

PIRs are made of a pyroelectric detector, which is an infrared-sensitive element. The sensor in motion detectors is wired as two halves to avoid measuring average IR levels. For a static object emitting IR, the produced voltages from the two halves cancel each other out. If one half detects more or less IR radiation than the other, a voltage will be produced to indicate a motion.

4.2.1.3.16 LiDAR LiDAR, also known as laser altimetry, is an acronym for light detection and ranging. It refers to a remote sensing technology that emits a focused light wave and calculates the time it takes for the reflected wave to be detected by the sensor in order to find ranges or distances. In theory, LiDAR is similar to the old radar (radio detecting and ranging) technology, except that it is based on discrete pulsing of laser. The object's coordinates are obtained from the time difference between the transmitted laser and the received, the angle at which the laser was transmitted, and the reference location of the sensor.

4.2.1.4 Wireless Sensors

A crucial aspect of the sensor system is the ability to provide means of transmitting the perceived information to an external processing unit or an actuator. Obviously, a wireless transmission provides mobility, portability, and convenience and enables the sensor to be deployed more flexibly.

Moreover, wireless sensors can be grouped together to form a network in order to provide a more sophisticated set of data and/or to communicate and exchange information. Sensors in a common network (wireless sensor network (WSN)) share data either through nodes that combine information at a gateway or where each sensor connects directly to gateways which act as bridges that connect the sensors to the Internet.

4.2.1.5 Multisensor Modules

Multisensor modules combine a wide range of sensors in addition to a limited-power processing unit, communication capability, cloud connectivity, and other peripherals. These modules are typically used as development platforms for the design and prototyping of IoT systems.

The Texas Instruments (TI) CC2650 SensorTag powered by a single coin cell battery, for example, is one of the most commonly used modules, and features the following components in a single package:

Sensor input
Ambient light sensor (TI Light Sensor OPT3001)
Infrared temperature sensor (TI Thermopile infrared TMP007)
Ambient temperature sensor (TI light sensor OPT3001)
Accelerometer (Invensense MPU-9250)
Gyroscope (Invensense MPU-9250)
Magnetometer (Bosch SensorTec BMP280)
Altimeter/Pressure sensor (Bosch SensorTec BMP280)
Humidity sensor (TI HDC1000)
MEMS microphone (Knowles SPH0641LU4H)
Magnetic sensor (Bosch SensorTec BMP280)
Push-button GPIOs
Reed relay (Meder MK24)
Output components
Buzzer/speaker
LEDs
Communications
Bluetooth Low Energy (Bluetooth Smart)
ZigBee
6LoWPAN

The module uses a processing module that includes an extremely low-power CPU (ARM Cortex M3) with a 128 KB of flash memory and 20 KB of static RAM (SRAM). While power-efficient, this limits the amount of processing and resources on this system. Typically, such limited-power devices will need to be supplemented by a gateway, router, smartphone, etc. Another module used for IoT prototyping is the PRISM introduced by Eleco with the following specification:

Communications: Bluetooth 4.0 (BLE)
Microcontroller: ARM Cortex-M0+ 32 bit

Power source voltage: +2.35–3.3 V
Consumption current: 5 mA (Peak current)
Standby current: 8 µA
Accelerometer: 3-axis ±2G (max. ±16G)
Compass: 3-axis ±1300 µT
Thermometer: −40 to +120 °C
Hygrometer (A hygrometer is a sensor used to measure humidity and water vapor in the atmosphere, in soil, or in confined spaces): 0–100%
Barometer: 300–1200 hPa
Illuminometer: 0–128 kLx
UV meter: UV index 0–11+

4.2.1.6 Signal Conditioning for Sensors

The voltages produced by analog sensors are extremely small (ranges from picovolts to millivolts). Thus, they need to be amplified before they can be used as an input to the analog-to-digital conversion stage. Such amplification is only one part of the signal conditioning process. Impedance matching, input-output isolation, and filtering may also be required before the signal can be processed and analyzed.

4.2.2 Actuators

Unlike sensors that provide information about a process/environment, actuators provide action. Actuators typically receive some type of control signal based on sensors' data that triggers a physical effect.

Actuators can also vary in type, function, and area of application. Some common categorizations are based on power, motion, and industry.

The most powerful use cases in IoT and wearable technology are those where sensors and actuators work together in an intelligent, complementing, and harmonious fashion. Such combination can be utilized to solve problems by simply elevating the data that sensors provide to actionable insight that can be acted on by work-producing actuators.[2] Examples of actuators include motors, relays,[3] speakers, and lights. Just like a sensor, an actuator may need a conditioning and/or driving circuit. Figure 4.2 depicts a signal flow in a sensor/actuator-based system.

2 **Haptics**, also known as kinaesthetic communication refers to the recreation of touch experience by applying forces, vibrations, or motions to the user using a variety of actuators.
3 **A relay** is a binary actuator that has two stable states, either latched (when energized) or unlatched when de-energized. The most popular relays are as follows: electromagnetic relays, which are constructed with electrical, mechanical and magnetic components, and have operating coil and mechanical contacts; solid-state relays, which use solid-state components to perform the switching mechanism without moving any parts; and hybrid relays.

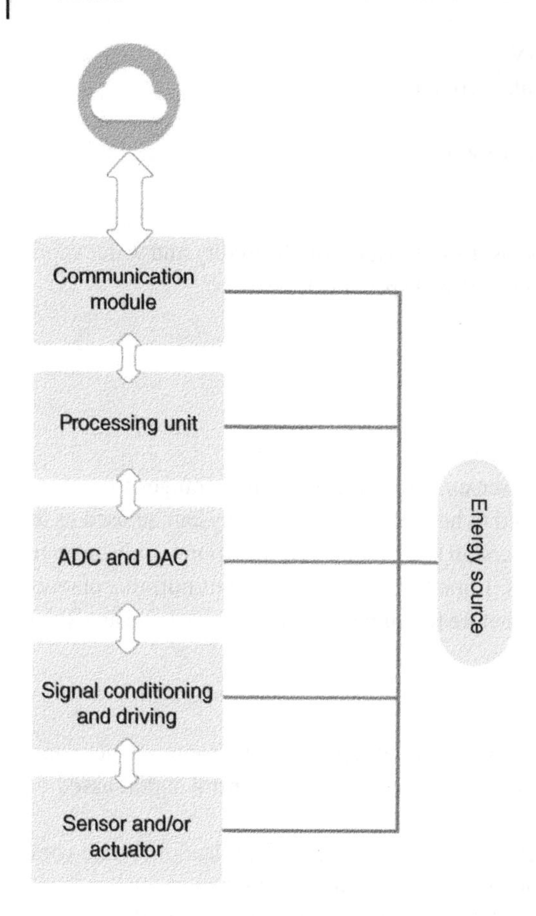

Figure 4.2 Sensor and actuator signal flow in a system.

4.2.3 Microcontrollers, Microprocessors, SoC, and Development Boards

The most essential component and what makes IoT and wearables a smart technology is either a microprocessor or a microcontroller.

A microprocessor incorporates the functions of a computer's central processing unit (CPU) in a single integrated circuit (IC) chip. It accepts digital data as input, processes it according to a sequence of instructions stored in the memory, and delivers results as output. Microprocessors are used in everything from the smallest handheld devices to the most powerful supercomputers. Microcontrollers, on the other hand, also have a RAM, ROM, and other peripherals integrated within a single chip along with the CPU.

The choice of the microprocessor in IoT and wearables is driven by the application, industry, and functions performed by the device. For most applications, a general purpose microprocessor unit would suffice; however, highly specialized devices would most likely require a dedicated application processor. It should be noted that nowadays microprocessor manufacturers incorporate most of the functions in a single chip which is crucial in reducing the overall size and cost of a wearable device. For example, the 32-bit ARM processor, which is a reduced instruction set computer (RISC) architecture developed by Advanced RISC Machines (ARM), is very common in IoT and wearables as it provides a sufficiently powerful computational performance and energy efficiency.

Another popular microcontroller is the Programmable System on Chip (PSoC)[4] developed by Cypress Semiconductor which integrates programmable analog and digital functionalities in a single chip utilizing the power of an ARM cortex-M core architecture.

It is also worth mentioning that some advanced devices have a separate coprocessor (sensor hub) dedicated to handle the sensors' data. This is critical when the device has a large amount of sensors data that needs to be processed in real time, which requires an uninterrupted CPU attention. This function is widely known as "sensor fusion."

It should also be noted that based on the features offered by the IoT or wearable product, the device may or may not require a specific operating system. For instance, a light-weight RTOS (real-time operating system) may be more than sufficient to operate a wristwatch that measures temperature, tracks a user's movement using a simple accelerometer, and displays time on a basic LCD display. On the other hand, a sophisticated smart watch that serves as an extension of a user's mobile phone needs to run an advanced operating system such as an iOS or Android.

A development board, on the other hand, is a prototyping solution that features a low-power CPU which typically supports various programming environments. The board, in essence, is a printed circuit board containing a microcontroller unit, interfacing circuitry, power management unit, and communication capability.

---

4 A **System on Chip (SoC)** refers to a grouping of all the components of an electronic system in a single-integrated circuit. In addition to the processing unit, memory, and bus, a SoC may contain a sensor(s), communication capability and other components that deal with data compression, data filtering, etc.

A supporting firmware in addition to data transfer to a cloud-based server is typically included.

Developing IoT and wearable applications is now more accessible with the growing availability of low-cost, off-the-shelf development boards, platforms, and prototyping kits. Such modular hardware offers greater flexibility to the designer.

While a microcontroller is a SoC that provides data processing and storage capabilities, a single board computer (SBC) is a step-up from microcontrollers. It allows the user to connect peripheral devices like keyboards and screens, in addition to offering more processing power and memory.

Sensors and actuators connect to the microcontroller and microcomputer through analog and digital general purpose input/output (GPIO) pins or through a bus. Standard communication protocols like SPI and I2C are used for in-device communication.

The ARM Cortex-M processors are very widely used in IoT and wearables. They support multiple clock and power domains. They also support advanced low-power techniques and provide different sleep modes. Below is a comparison between different M processors:

Cortex-M0 processor: The smallest ARM processor which makes it ideal for low-cost microcontrollers for general data handling and simple input–output control tasks.

Cortex-M0+ processor: The most energy-efficient ARM processor. It features the same instruction set as Cortex-M0; in addition to IoT and wearable application, it is suitable for general data handling and input–output operations.

Cortex-M3 processor: This one is by far the most popular ARM processor used in high-performance microcontrollers that are also energy efficient.

Cortex-M4 processor: This processor features all the functions of the Cortex-M3 processor, with additional instructions to support Digital Signal Processing (DSP) operations.

Cortex-M7 processor: The highest performance processor with open memory interface options and a more powerful DSP performance.

Cortex-M23 processor: This processor is similar to the Cortex-M0+ processor, but features a newer architecture version called ARMv8-M, which adds a security extension and several additional instructions.

Cortex-M33 processor: This processor is similar to the Cortex-M3/M4 processor but also supports ARMv8-M architecture in addition to enhanced system level features.

FPGA (field-programmable gate arrays), on the other hand, are integrated circuits which are groups of programmable logic gates, memory, and other elements. FPGAs could be coupled with a processor to interface with the outside world, to provide lowest power, lowest latency and best determinism, and to leverage more

Figure 4.3 The arm MPS2+FPGA prototyping board. The platform offers a large FPGA for prototyping Cortex-M based designs with a range of debug options. *Source*: Photo courtesy of ARM.

advanced software functions such as web services or security packages. Figure 4.3 shows a typical FPGA development board.

4.2.3.1 Selecting the Right Processing Unit for Your IoT or Wearable Device

There are a plethora of development boards, microcontrollers, and microprocessors available in the market and selecting the right one is dependent on a number of factors and is highly bounded by the targeted application. Below are some important factors the product developer should be aware of:

- **Compatibility:** Does the unit support the sensors and actuators required in your project?
- **I/O Support:** The number of Input/Output ports determines the number of sensors and actuators that can be used in the project.
- **Architecture**: Can the architecture handle the complexity of your project? Selecting the right architecture depends on the functional requirements of your project and how much computing power your application will need.

- **Clock Speed and Memory:** Is the processing unit equipped with adequate memory for your project? Also, some IoT or wearable applications will run adequately at low speeds, some will run the processor at higher speed to achieve a more demanding task, and some may have different clock needs depending on the dynamics of the application. The designer needs to make an informed decision concerning this before prototyping.
- **Power Requirement:** How much power will the unit need? What is its power consumption while in action and during idle time? Energy efficiency is extremely important for wearables and mobile/portable (nonwired) IoT applications.
- **Customer and Community Support:** Is good documentation for your unit available? Is customer support reputable and reliable? This is crucial when it comes to making informed decisions on how to professionally use your unit.
- **Add-On Capabilities:** Some wearable and IoT applications may require DSP capability for analysis and modification of signals. Hence, a dedicated digital signal processor may be required onboard.
- **Connectivity:** IoT and most wearables must have a form of connectivity. The availability of connectivity type(s), such as Ethernet, WLAN, and BLE, needed for the project on board is a great advantage.
- **Security:** Security is of paramount importance in IoT and wearable devices. Hardware support for security may be required or preferred in some applications.

Because IoT and wearable technology cover a wide spectrum of applications, processing and wireless requirements vary drastically. For example, some wearable devices perform a small amount of processing and merely upload data to the cloud. Such devices use low-cost, low-power microcontrollers. The wireless connectivity is typically integrated within the board. Other devices, such as smart watches and security cameras, require upper-scale processors for data analysis or driving a display.

It is also worth noting that many smart devices in the market today use repurposed smartphone processors. However, other companies have gone the extra mile and designed processors dedicated for IoT devices and wearables.

4.2.4 Wireless Connectivity Unit

Clearly, wireless connectivity is of paramount importance in IoT and wearable devices as most of them need to interact with a networking device. Supporting one or more wireless communication protocols such as Wi-Fi, BLE, and IEEE 802.15.4 LR-WPAN (Low Rate Wireless Personal Area Network) is typically required in these devices.

Obviously, no wireless transmission is possible without an antenna, and thus, the functionality and efficiency of any IoT or wearable device with a wireless connectivity are primarily dependent on the properties of the integrated antenna unit.[5]

In general, the nature of small-form factor IoT and wearable devices requires the integrated wireless connectivity components to be compact, light-weight, low-profile, and mechanically robust, simultaneously. They also must exhibit reliability, high efficiency, and desirable radiation characteristics.

In wearable technology, there are a number of additional challenges that engineers face when designing antennas and wireless systems that do not exist in conventional wireless units which will be discussed in Chapter 6.

Designers must choose between laying out their own RF transceiver chip, antenna, and impedance matching circuit in the form of a chip or a PCB or going for off-the-shelf RF modules. Designing a dedicated PCB that meets the requirements of electromagnetic compatibility (EMC), electromagnetic interference (EMI), and regulations could be a very lengthy and expensive process.

The commercially available system-in-package (SiP) RF modules integrate all the necessary components, including the antenna, and they typically come pretested and precertified. This minimizes a lot of design complexity and reduces development time, energy, and risks, which allow developers to focus on their target applications. Figures 4.4 and 4.5 show a printed monopole antenna and typical radiation patterns of monopole/dipole, and microstrip antennas, respectively.

4.2.5 Battery Technology

When designing an IoT or a wearable product, it is important to consider throughout the design process how the performance of the device will affect its energy budget. Energy consumption, battery capacity, and duty cycles are among the key components of the energy budget.

Although computer architects are thriving to produce ultra-low-power microprocessors and microcontrollers, the power demand is still high due to a larger size, higher resolution displays, and multiple apps that are in use simultaneously. Unfortunately, there is no version of Moore's law that applies to batteries as the annual improvement rate in battery capacity does not exceed 8%.

---

5 **Antennas** are electromagnetic radiators that convert electrical currents to electromagnetic waves at the transmitting end and from electromagnetic back to electrical currents on the receiving end. The most common types of antennas in terms of radiation are omni-directional and directional antennas. Omni-directional antennas radiate its energy in all directions equally except top and bottom (donut-shaped radiation pattern), whereas a directional antenna will focus its energy in a certain direction. Common antennas in IoT and wearable devices are as follows: chip, PCB, and wire antennas.

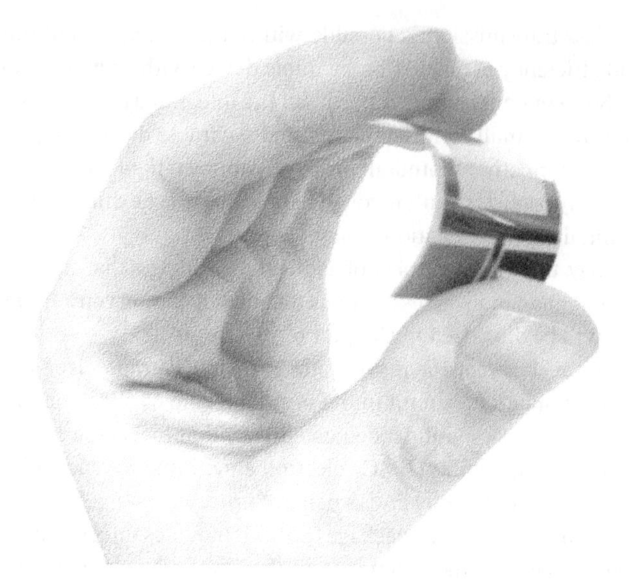

Figure 4.4 Printed monopole antenna intended for integration within flexible electronics.

According to a recent study, one-third of Americans who have a wearable device stop using it within the first six months due to battery life limitation.[6]

Lithium-based rechargeable batteries have become the obvious choice in handheld applications since its commercial debut in the early 1990s. There are several reasons attributed to its dominance: higher cell voltage compared to Ni-based batteries, lighter weight, higher energy density, relatively simpler manufacturing process, and higher recharge-ability rate. It should be noted that the principal difference between lithium ion and lithium polymer (the main Li-based battery technologies) is that lithium ion has a higher capacity, whereas the lithium polymer is lighter in weight.

On the other hand, both flexible and printed batteries offer promising compact solutions for wearable devices such as in transdermal drug delivery patches, temperature sensors, and RFID tags. These include polymeric lithium, solid-state, printed zinc-based batteries, in addition to flexible supercapacitors.

---

6 **Moore's law** refers to an observation pointed out by Gordon Moore (a co-founder of Intel Corp.). His observation concludes that the number of transistors per square inch in integrated circuits is doubled each year. Although the rate seemed to hold true from 1975 until around 2012, the rate started to slow in 2013, and in 2015, Gordon Moore himself stated that the growth rate would reach saturation in the following decade.

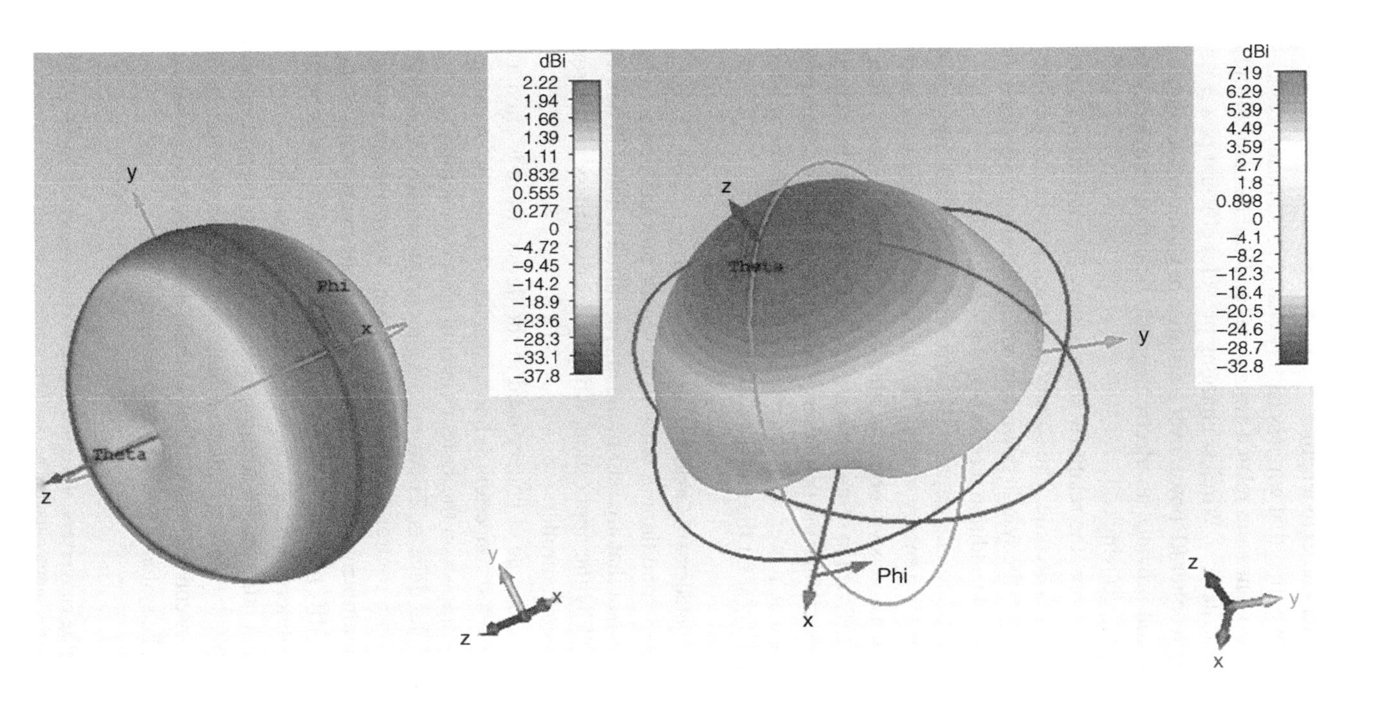

Figure 4.5 Omni directional radiation pattern (left), and semi-directional (hemi-spherical radiation pattern (right).

Another compact option is using a miniaturized packaging of traditional batteries, as in the battery offered by Panasonic that is available in a cylindrical package of 3.5 mm diameter and 2 cm length.

It is also worth noting that wireless battery charging is already available in the market for smartphones and related accessories and will be naturally adopted by IoT and wearable devices. Wireless charging provides efficient power transfer to batteries through either RF power over a distance or via inductive coupling where a transmitting coil provides an electromagnetic field that transfers energy to a closely positioned receiving coil.

In some applications, the relatively short battery life adds significant cost to a given system over its lifetime. The battery unit itself may be inexpensive, but the costs associated with the system downtime when the battery is drained (or during replacement) can add additional costs over a product's lifecycle.

Solar energy is potentially capable of harvesting significantly more power than many other alternative energy sources like thermoelectric and piezoelectric transducers, electrodynamic switches, and ambient RF signals. Hence, integrating a solar panel into IoT, and in some applications, wearables, could be a tangible solution. Solar panels can harvest light energy both indoors and outdoors providing a source of consistent power, thus increasing the lifetime of the product while reducing the total cost for the end user.

4.2.5.1 Power Management Circuits

Embedding signal conditioning within the sensor provides some considerable advantages. The data that are sent to the microcontroller unit will be swiftly and easily interpreted by the application, which gives rise to less power being consumed by the microcontroller.

Depending on the type of battery used in the device, there is often a requirement for step-up boost converters or boost-switching regulators.[7] A careful choice can make a huge impact on the system's overall power consumption.

For more complex devices, a power management integrated circuit (PMIC) provides a more precise control over the entire system. From a single power source, one can derive multiple voltage rails to power different components of an embedded system. A PMIC may also offer additional functionality for general system control, such as timers, voltage sequencing, and reset capability.

It is also worth noting that in addition to using low-power semiconductor components, utilizing software techniques, including stacks, encryption, and data processing, are key considerations. Each of these design factors can have a significant impact on the system's overall power budget.

---

7 **Step-up switching converters**, also called boost switching regulators, enable a higher voltage output than the input voltage. The output is regulated, as long as the power draw is within the specified output power.

4.2.6 Displays and Other User Interface Elements

How a user interacts with an IoT or wearable device is an essential design aspect. Complexity should be minimized, and the interactive experience should be as intuitive as possible.

Inherited from mobile phones and handheld electronics, displays with capacitive touch screen capability are the obvious choice in providing a user with a feedback and/or a user interface platform.

The emergence of flexible and curved displays is considered as the major breakthrough in wearable display technology, and their introduction to the electronics market was certainly driven by wearables.

Organic light-emitting diode (OLED)-based displays, which emits light from an organic compound layer in response to electric current, has substantial performance advantages in wearables. By only powering the active pixels, a considerable amount of energy can be saved. Furthermore, OLED offers a borderless and semitransparent projection which gives rise to an improved user experience especially in smart glasses applications.

Another noteworthy example of wearable technology-driven development in display technology is the digital light processing (DLP). It has many performance gains over traditional Liquid-crystal display (LCD)-based projectors since it enables noise-free, precise image quality, and color reproduction capability.

In addition to touch screens, buttons, switches, knobs, and sliders, there are other ways a user can electronically interact with the smart device. Other important elements used in IoT and wearables include buzzers and vibrating motors which are essential in alerting the user when certain activities take place. For instance, the vibrating motor in smart watches is utilized to alert a user when a message is received. LEDs and digital segment displays are also very commonly used in wearables to provide feedback.

4.2.7 Microphones and Speakers

Many IoT devices and wearables have integrated microphone and speaker to perform voice commands through a user interface platform. Different types of microphones and speakers can be embedded in these devices including piezoelectric MEMS microphones and microspeakers.

4.3 Conclusion

Advancement across the various disciplines of electrical engineering offers unique advantages and opportunities to interact with and influence our environment. This is the basis of IoT and wearable technology, and it opens up a world of novel and innovative possibilities. Embedding sensors and/or actuators within ordinary objects and networking them is a great way to enable advanced and well-coordinated automations that improves efficiency, saves costs, and convenience.

This Chapter introduced sensors and actuators and their characteristics. It also included descriptions and practical examples of microprocessors and basic guidelines to choose the right one for a given application. Additionally, discussions on wireless communication modules, energy sources and management circuits, and various other peripheral components in IoT and wearable devices were provided.

Problems

1. Based on the anatomy of a general connected device depicted in Figure 4.1, sketch a similar diagram pertaining a smart watch.

2. Based on the anatomy of a general connected device depicted in Figure 4.1, sketch a similar diagram pertaining a smart camera-based smart lock and doorbell system.

3. Pick a wearable or IoT device of your choice, then list all of the device's components (external and internal).

4. What is a MEMS sensor? Research five examples from the literature and compare between their mechanisms of operation.

5. What are the different types of accelerometers? How would you characterize a typical one?

6. What are some of the most common types of thermocouples?

7. Research the most common types of motors used in IoT and wearable applications.

8. What would be a good choice of an antenna topology for a fitness tracker? Why?

9. What would be a good microprocessor/microcontroller choice for a wearable device that makes one heart rate reading every six hours? Justify your choice.

10. You are tasked to prototype a virtual home assistant (i.e. similar to Amazon Echo Dot). Make a list of all the tasks needed to create such a device, along with a list of all the components needed based on what you have learned in this chapter.

Technical Interview Questions

1. What is the difference between microprocessors, microcomputers, and microcontrollers?
2. What are some of the common errors found in embedded systems?
3. What is the difference between an active and a passive transducer?
4. What are the elements of a conditioning circuit? Why is it needed in sensing systems?
5. What is a PMIC? Why is it needed in IoT and wearable devices?
6. Draw a functional block diagram of a sensing system.
7. In antennas, what is a return loss and VSWR? How are they related?
8. Sketch a physical design and PCB with all the necessary components for a smart light bulb.
9. You are given a PCB, explain the potential sources of noise.
10. Which type of technology would you choose for a touch screen on a connected device?
11. What is a nonlinear phase filter? Derive group delay for this filter.
12. There is a new sensor being developed. What is your approach to test it?
13. What are the electrical and mechanical properties of flex cables?
14. How is a pressure sensor fabricated?
15. What is the difference between TTL and CMOS?

Further Reading

Aasin Rukshna, R., Anusha, S., Bhuvaneswarri, E., and Devashena, T. (2015). Interfacing of proximity sensor with My-RIO toolkit using labVIEW. *International Journal for Scientific Research & Development* 3 (01): 562–566.

Ahson, S.A. and Ilyas, M. (2011). *Near Field Communications Handbook (Internet and Communications).* Boca Raton, FL: CRC Press Taylor and Francis, 23 September). ISBN-10: 1420088149.

Atzori, L., Iera, A., and Morabito, G. (2010). The internet of things: a survey. *Computer Networks* 54 (15): 2787–2805.

Azuma, R., Behringer, R., Feiner, S. et al. (2001). Recent advances in augmented reality. *Computers & Graphics* 21: 34–47.

Banaee, H., Ahmed, M.U., and Loutfi, A. (2013). Data mining for wearable sensors in health monitoring systems: a review of recent trends and challenges. *Sensors* 13 (12): 17472–17500.

Bennett, T.R., Jafari, R., and Gans, N. (2014). Motion based acceleration correction for improved sensor orientation estimates. *2014 11th International Conference on Wearable and Implantable Body Sensor Networks*, Zurich, Switzerland.

Bierl, L. (1996). *Precise Measurements with the MSP430.* Dallas, TX: Texas Instruments.

Bilal, M. (2017). "A review of internet of things architecture", technologies and analysis smartphone-based attacks against 3D printers. arXiv preprint arXiv:1708.04560, 1–21.

Botterman, M. (2009). For the European commission information society and media directorate general, networked enterprise & RFID unit – D4. *Internet of Things: An Early Reality of the Future Internet, Report of the Internet of Things Workshop*, Prague.

Cox, D. (2012). Implementing Ohmmeter/Temperature Sensor; Microchip Technology AN512: Chandler, AZ, USA, 1994. J. Low Power Electron. Appl., 2 280.

Crabtree, G., Kocs, E., and Trahey, L. (2015). The energystorage frontier: lithium-ion batteries and beyond. *Materials Research Society MRS BULLETIN* 40: 1067–1076.

Dietz, P.H., Leigh, D., and Yerazunis, W.S. (2002). Wireless liquid level sensing for restaurant applications. *Proceedings of The 1st IEEE International Conference on Sensors*, Orlando, FL, USA (12–14 June 2002), pp. 715–719.

Fayyad, U., Piatetsky-Shapiro, G., and Smyth, P. (1996). From data mining to knowledge discovery in databases. *American Association for Artificial Intelligence* 17: 117–152. 0738-4602-1996.

Gaitán-Pitre, J.E., Gasulla, M., and Pallàs-Areny, R. (2009). Analysis of a direct interface circuit for capacitive sensors. *IEEE Transactions on Instrumentation and Measurement* 58: 2931–2937.

Guth, J., Breitenbucher, U., Falkenthal, M. et al. (2016). Comparison of IoT platform architectures: a field study based on a reference architecture. *Cloudification of the Internet of Things (CIoT)*, Paris (23–25 November 2016).

Hanes, D. (2017). *IoT Fundamentals: Networking Technologies, Protocols, and Use Cases for the Internet of Things*, 1e. Indianapolis, IN: Cisco Press.

Ho, E., Jacobs, T., Meissner, S. et al. (2013). ARM testimonials. In: *Enabling Things to Talk* (eds. A. Bassi, M. Bauer, M. Fiedler, et al.), 279–322. Berlin, Heidelberg: Springer.

ARM.com (2020). Arm Cortex-M series processors (guide). https://developer.arm.com/products/processors/cortex-m (accessed March 2020).

Huising, J.H. (2008). Smart sensor systems: Why? Where? How? In: *Smart Sensor Systems* (ed. G.C.M. Meijer), 1–21. Chichester, UK: Wiley. 3.

Kang Lee, F.J. and Lanctot, P. (2017). *Internet of Things: Wireless Sensor Networks*. Geneva, Switzerland: International Electrotechnical Commission.

Khaleel, H.R., Al-Rizzo, H.M., Rucker, D.G., and Mohan, S. (2012). A compact polyimide-based UWB antenna for flexible electronics. *IEEE Antennas and Wireless Propagation Letters* 11: 564–567.

Lee, B.M. and Ouyang, J. (2014). Intelligent healthcare service by using collaborations IOT personal health device. *International Journal of Bio Science and BioTechnology* 6 (1): 155–164.

Mohammadi, M., Aledhari, M., and Al-Fuqaha, A. (2015). Internet of things: a survey on enabling technologies, protocols and applications. *IEEE Communications Surveys & Tutorials* 17 (4): 2347–2376.

Mukherjee, M., Adhikary, I., Mondal, S. et al. (2017). A vision of IoT: applications challenges and opportunities with Dehradun perspective. *Advances in Intelligent Systems and Computing* 479 (4): 553–559.

Ngu, Q.Z.S.A.H., Gutierrez, M., Metsis, V., and Nepal, S. (2017). IoT middleware: a survey on issues and enabling technologies. *IEEE Internet of Things Journal* 4: 1–20.

Nitzan, M., Romem, A., and Koppel, R. (2014). Pulse oximetry: fundamentals and technology update. *Medical Devices* 7: P231–P239.

Pallàs-Areny, R. and Webster, J.G. (2001). *Sensors and Signal Conditioning*, 2e. New York: John Wiley & Sons.

Reverter, F. and Pallàs-Areny, R. (2005). *Direct Sensor-to-Microcontroller Interface Circuits. Design and Characterization*. Barcelona, Spain: Marcombo.

Reverter, F., Gasulla, M., and Pallàs-Areny, R. (2007). Analysis of power-supply interference effects on direct sensor-to-microcontroller interfaces. *IEEE Transactions on Instrumentation and Measurement* 56: 171–177.

Richey, R. (1997). Resistance and Capacitance Meter Using a PIC16C622; Microchip Technology AN611: Chandler, AZ, USA.

Riva, G., Mantovani, F., and There, B. (2012). *Understanding the Feeling of Presence in a Synthetic Environment and Its Potential for Clinical Change*. London, UK: Riva and Mantovani, Intech.

Sadowsky, G., Dempsey, J.X., Greenberg, A. et al. (2003). *Information Technology Security Handbook*. Washington, DC: World Bank.

Texas Instruments (2014). *Analog Front End (AFE) for Sensing Temperature in Smart Grid Applications Using RTD (Handbook)*. Dallas, TX: Texas Instruments.

Toma, I., Simperl, E., and Hench, G. (2009). A joint roadmap for semantic technologies and the internet of things. *Proceedings of the Third STI Road mapping Workshop*, Crete.

Varghese, B. and Buyya, R. (2018). Next generation cloud computing: new trends and research directions. *Future Generation Computational Systems* 79: 849–861.

Woodard, J., Weinstock, J., and Lesher, N. (2014). *Integrating Mobiles into Development Projects*. Washington, DC: USAID.

Xia, F. (2009). Wireless sensor technologies and applications. *Sensors* 9 (11): 8824–8830.

Yamanaka, K., Vestergaard, M.'d.C., and Tamiya, E. (2016). Printable electrochemical biosensors: a focus on screen-printed electrodes and their application. *Sensors* 16 (10): 1761.

5

Communication Protocols and Technologies

5.1 Introduction

IoT and most wearable devices must connect to a network for their data to be transported and utilized. In addition to the wide range of components that make up these devices, there are also several communication technologies and protocols used to connect them.

Protocols ensure that data from one device or sensor are reliably and securely delivered and understood by another device or system. Given the diverse array of IoT and wearable devices available, using the right protocol in the right context is of paramount importance.

There exists an overwhelming number of connectivity options for designers working on products and systems for IoT and wearable technology. How protocols and standards support secure and reliable data exchange in the ecosystem is a question that any serious designer should know the answer to. It is important to take into account the application requirements, architecture, and factors that impact signal quality, bandwidth, and range.

This Chapter takes a look at the characteristics and basics of the communication protocols that IoT and wearables employ for their data exchange, along with a dive into some of the most common technologies being deployed today.

5.2 Types of Networks

As we saw in the previous chapters, IoT and wearable system in general use a three-layer architecture: devices, gateways, and data centers/cloud. The communication can be between the devices themselves, a device and gateway, a gateway to a data center, or between data centers.

Fundamentals of IoT and Wearable Technology Design, First Edition. Haider Raad.

Published 2021 by John Wiley & Sons, Inc.

Computer networks are typically categorized based on the range they provide. The size of a network in IoT and wearables can vary from connecting two devices within the user's body to two smart systems communicating across the world. Below is a list of the most popular types of networks used in IoT and wearable technology:

Body Area Network (BAN): A network that connects devices within or inside the user's body such as wearables, insertables, and implants. This type of network is also known as a wireless body area network (WBAN), body sensor network (BSN), or medical body area network (MBAN).

Personal Area Network (PAN): A network that connects devices within a room or a radius of a person's workspace such as wearables, laptops, and personal gadgets powered by Bluetooth and ZigBee communication protocols. A wireless PAN (WPAN) covers anywhere between a few inches and a about 30 ft.

Local Area Network (LAN): A network that connects devices within a premise or building. Ethernet and Wi-Fi are LAN's two most common technologies.

Campus/Corporate Area Network (CAN): A network that combines smaller local area networks within a limited geographical area such as a university, school district, or an enterprise.

Metropolitan Area Network (MAN): A network that connects multiple LANs within a metro city into a bigger network.

WAN (Wide Area Network): A network that integrates multiple LANs and MANs into a single large network laid out across a wide geographical area such as a country.

Figure 5.1 depicts the major types of networks.

An example of a LAN setup is depicted below in Figure 5.2.

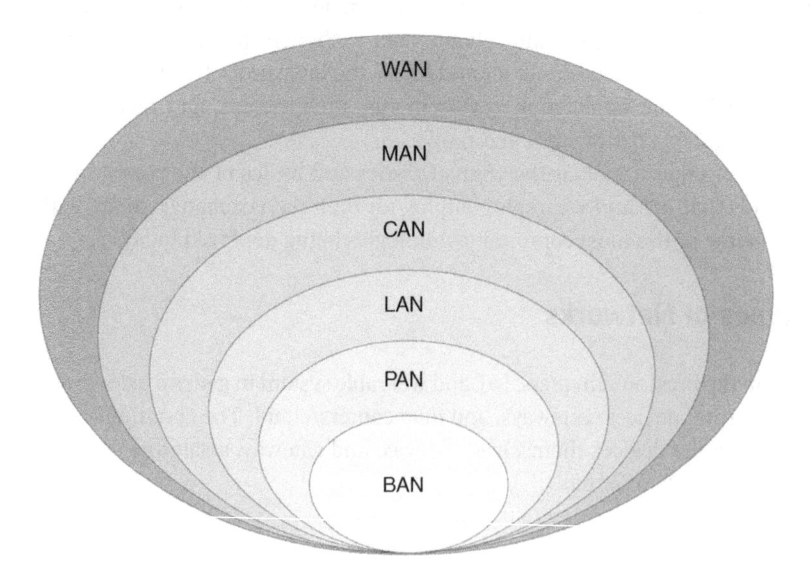

Figure 5.1 Types of computer networks.

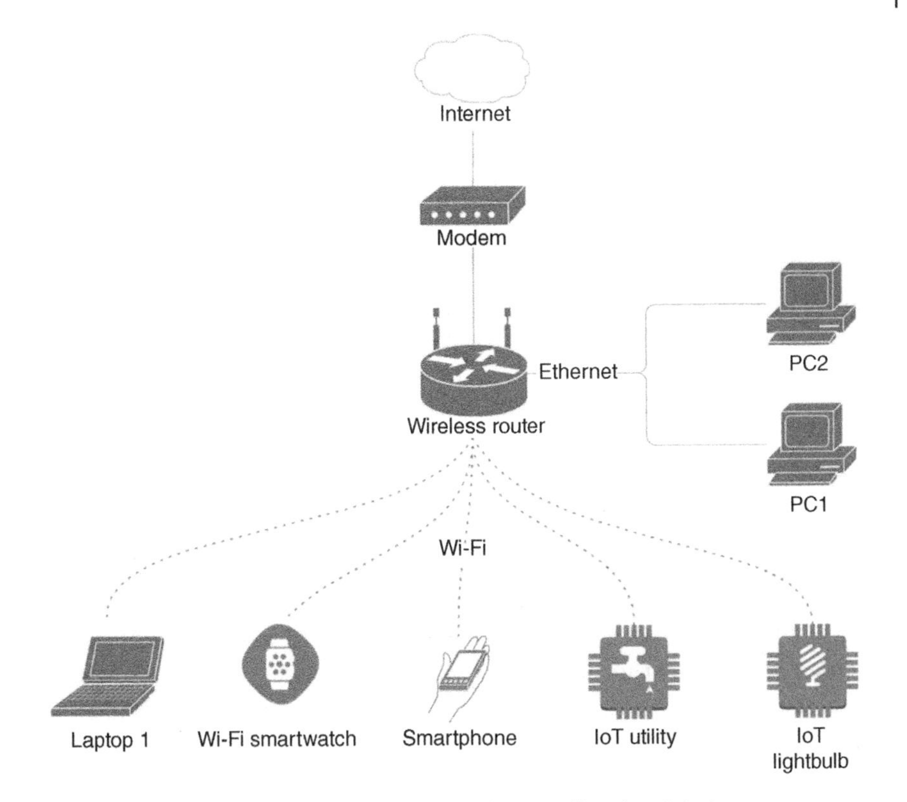

Figure 5.2 An example of a local area network with a diversity of devices.

5.3 Network Topologies

Networks can also be categorized according to their connectivity configuration, i.e. topology. A network topology refers to how computers, printers, and other devices are connected, and describes the layout and routing paths of such connections in a given network. When referring to topology in the context of IoT and wearable technology, it describes how sensors, actuators, gateways, and servers communicate with one another. There are a number of common topologies used in the field of IoT and wearable technologies, including point to point, mesh, bus, ring, and star.

5.3.1 Mesh

Mesh topology is a type of connection where all nodes work together to distribute data in a network. The primary advantage of this topology is that it uses low power and shorter links (typically less than 100 ft), which promotes a longer battery life.

The main disadvantage, however, is that if one node goes down, an entire piece of the network can fail due to the interconnected nature of mesh networks. This topology is commonly found in IoT deployments and typically used in home automation and smart buildings utilizing protocols such as ZigBee, Thread, and Z-Wave.

5.3.2 Star

The star network is one of the most common network topologies. In this topology, every host is connected to a central hub that acts as a conduit to transmit messages. An advantage of star topology is that all the complexity in the network is driven to a central node, which also introduces a single point of failure.

5.3.3 Bus

Bus topology is a simple and easy to install network configuration in which each node is connected to a common cable (bus). Data is transmitted in either direction along the bus until it reaches its intended destination. However, there is a single point of failure: If the bus fails, the entire network fails. This topology can be used in sensor networks that are physically wired together.

5.3.4 Ring

A ring topology, which is not common in IoT and wearable technology but worth mentioning here, is technically a bus topology arranged in a closed loop where data are transmitted around the ring in one direction. When one node passes data to a destination, the data have to go through each intermediate node. These intermediate nodes retransmit the data (acting as a data repeater) which keep the signal strong over a long distance.

5.3.5 Point to Point

A point-to-point network topology is very common in wearable technology where a direct connection is established between two nodes in a network. An example of this type of network is a Bluetooth link between a smartphone and a fitness tracker.

Low cost and simplicity are the main advantages of point-to-point topology; however, the network is not scalable beyond these two nodes; therefore, it is not suitable for typical IoT applications where a multitude of nodes (i.e. sensors, smart objects) exists in the network and in need of communicating with each other. Figure 5.3 shows some common types of network topologies.

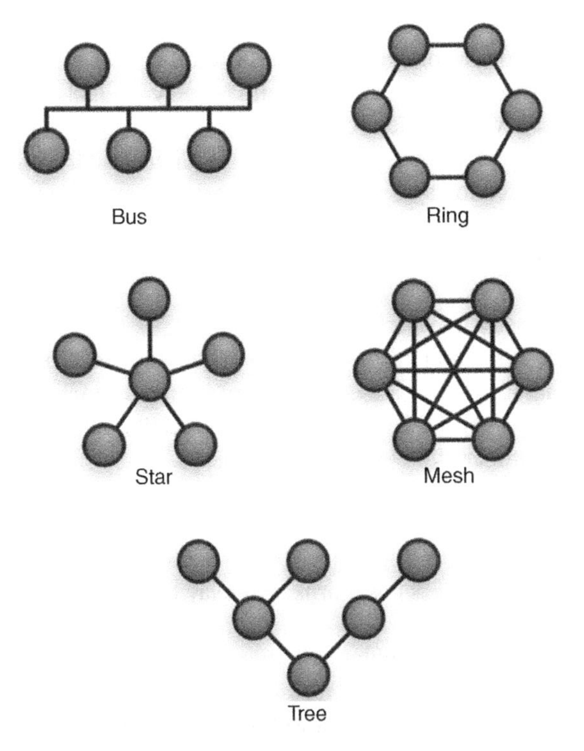

Figure 5.3 Common types of network topologies.

5.4 Protocols

5.4.1 Application Layer Protocols

The reader may ask: why are there other protocols outside of HTTP to transport data across a network? HTTP, the underlying protocol used by the World Wide Web, has provided outstanding services and abilities for the Internet in the past three decades, yet it was designed and aimed for general purpose computing in client/server models. IoT and wearable devices, on the other hand, can be very constrained in terms of power and bandwidth, which triggers the need for a more efficient, secure, and scalable protocols for managing different types of devices communicating via different network topologies.

The application layer serves as the interface between the device and the end user. Below are some widely used application layer protocols:

5.4.1.1 Constrained Application Protocol (CoAP)

CoAP is an application layer protocol designed for use in resource-constrained Internet devices. While the existing Internet infrastructure is freely available and can be used by any connected device, it often renders too heavy and power-consuming for most applications in IoT and wearable technology. Conceived to simplify the integration of HTTP-based IoT systems with the web, CoAP relies on the User Datagram Protocol (UDP) for ensuring secure and low overhead communication between network nodes.

5.4.1.2 Message Queuing Telemetry Transport (MQTT)

Probably the most widely adopted standard in the realm of IoT and wearables. MQTT is a lightweight messaging protocol aimed for battery-operated connected devices; it has been especially designed for unreliable communication networks in order to serve the growing number of low-power smart objects.

MQTT is based on subscriber, publisher, and broker model. As mentioned in Chapter 3, the sensor can be set up to be a publisher (publishes the required information), the application can be set as the subscriber (receives the information), and an intermediate system is set up to act as a broker to relay the information between the publisher and the subscriber.

Despite its advantages, MQTT can be problematic in some applications due to the dependency on TCP for transmission of messages and managing long topic names. Moreover, MQTT does not support a well-defined data representation and a structure model for device management, which leaves the implementation of its capabilities for data and device management platform to be vendor-specific.

5.4.1.3 Extensible Messaging and Presence Protocol (XMPP)

XMPP is an open standard and is based on Extensible Markup Language (XML) for real-time communication with proven scalability and security. XMPP has been around for a while, and before its use in IoT, it has been used widely in other applications including Voice over IP (VoIP) and instant messaging.

XMPP is accessible through any programming language which makes it very flexible; however, it has a larger overhead compared to lightweight protocols such as MQTT. Another drawback of XMPP is that it does not offer quality of service (QoS) or end-to-end encryption.

5.4.1.4 Data Distribution Service (DDS)

DDS is a peer-to-peer protocol developed by the Object Management Group (OMG) for real-time, scalable, and high-performance communications in IoT and M2M. DDS aims at simplifying deployment, increasing reliability, and reducing complexity.

5.4.1.5 AMQP (Advanced Message Queuing Protocol)

AMPQ is an open application layer protocol designed to enable interoperability in message-oriented middleware environments.

5.4.2 Transport Layer Protocols

This layer is responsible for an end-to-end communication over a network across multiple layers.

5.4.2.1 Transmission Control Protocol (TCP)

TCP is the dominant protocol in the Internet world. It breaks down large sets of data into packets that can be forwarded and reassembled as needed.

5.4.2.2 User Datagram Protocol (UDP)

UDP is used primarily for establishing low-latency and lossless connections over the network.

5.4.3 Network Layer Protocols

The network layer provides data routing paths for network communication and is responsible for packet forwarding including routing through intermediate routers.

5.4.3.1 IPv4 and IPv6

IPv4 and IPv6 are the two major versions of the Internet Protocol responsible for the delivery of data packets between computing nodes over the Internet and for uniquely identifying these nodes using IP addresses. Data packets include headers, which are metadata related to the data packet in addition to the actual data itself. The metadata includes important information such as sender and recipient IP addresses.

Due to the exponential increase in connected devices, we started to run out of IPv4 addresses. The solution is for IPv6 with a 128-bit destination address size to accommodate this increase along with improved routing and security.

5.4.3.2 6LoWPAN

6LoWPAN stands for IPV6 over low-power wireless personal area networks. It's a standard aimed at enabling battery-operated IoT and wearable devices, which often operate under constrained power budget to communicate using IPV6 packets. 6LoWPAN uses header compression and other power saving techniques allowing devices to communicate over IEEE 802.15.4 networks which defines the operation of low-rate WPANs.

5.4.3.3 RPL

RPL is a routing protocol designed for low power and networks with low-power devices which may experience packet loss (lossy networks). RPL is optimized for multi-hop and many-to-one communication, but also supports one-to-one messages.

RPL protocol is typically implemented in wireless sensor networks, and the most used operating system for its realization is Contiki which is an open-source OS developed for use in lower-end computers and microcontrollers.

5.4.3.4 Thread

A relatively new IP-based open-standard networking protocol, Thread, is a low-power wireless mesh networking protocol designed for smart home applications. Thread allows IoT devices to connect directly to the cloud with improved security and reliability.

5.4.3.5 LoRaWAN

LoRaWAN (long-range wide area network) is a network protocol designed with low power IoT devices in mind. LoRaWAN is a medium access control (MAC) layer (a sublayer of the data link layer) protocol but acts mainly as a network layer protocol for managing communication between gateways and end-point devices.

Stacks for the Web and connected devices are shown in Figure 5.4.

5.4.4 Protocols and Technologies in Physical and Data Link Layers

The physical and data link layers comprise devices and physical networks connecting them with other devices, network, and/or gateways. When designing

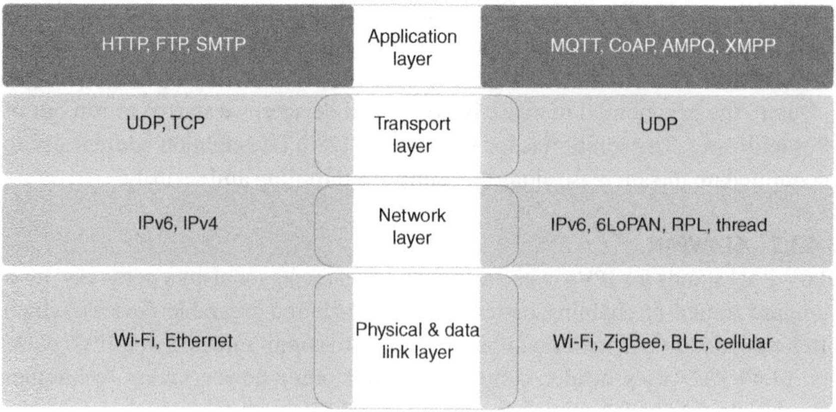

HTTP, FTP, SMTP	Application layer	MQTT, CoAP, AMPQ, XMPP
UDP, TCP	Transport layer	UDP
IPv6, IPv4	Network layer	IPv6, 6LoPAN, RPL, thread
Wi-Fi, Ethernet	Physical & data link layer	Wi-Fi, ZigBee, BLE, cellular

Figure 5.4 Typical TCP/IP stack (left), IoT/wearable devices protocol stack (right).

a new connected product, there are a bewildering number of protocols, standards, and technologies to choose from.

In a perfect world, networks would consume extremely small amount of power, offer a very wide range of coverage, and have a very large bandwidth. Unfortunately, this does not exist at the moment.

The available connectivity options are all governed by a trade-off between power consumption, range, and bandwidth (data rate). Without focusing on wired technologies[1], below are some of the popular physical and data link layer wireless protocols and technologies:

5.4.4.1 Short Range

5.4.4.1.1 Bluetooth (Short Range, High Data Rate, Low Power) Bluetooth is a short-range wireless communications technology standard that can be found in most smartphones and portable devices, which offers a major advantage for wearables and other personal gadgets.

Bluetooth has been a well-known technology for a long time, but not long ago a new WPAN technology introduced by the Bluetooth Special Interest Group (Bluetooth SIG) aimed at novel applications in the healthcare, wearables, security, and home entertainment industries. Compared to the legacy Bluetooth, Bluetooth Low Energy (BLE), formerly known as Bluetooth Smart, provides considerably reduced power consumption and cost while maintaining a comparable range.

Bluetooth can run various applications over different protocol stacks; however, each one of these stacks uses the same Bluetooth link and physical layers.

5.4.4.1.2 NFC and RFID (Short Range, Low Data Rate, Low Power) NFC (Near Field Communication) offers a set of communication protocols and technologies using electromagnetic fields that enable simple and secure two-way interaction between electronic devices.

NFC has its origins in radio-frequency identification (RFID) technology, which uses electromagnetic radiation to encode and receive information. Any NFC-enabled device has a microchip that is activated when it gets in close proximity to another NFC-enabled device (typically less than10 cm).

NFC solves many of the challenges associated with IoT and wearable devices such as offering a simple tap-and-go mechanism which makes it easy and intuitive to connect two different devices. Also, the short range feature of NFC prevents against unauthorized access.

---

1 **Ethernet** for IoT is a simple and inexpensive wired connection solution that provides fast data connection and low latency in stationary IoT applications.

5.4.4.1.3 Z-Wave (Short Range, Low Data Rate, Low Power) Z-Wave is a low-power wireless communications technology that is primarily designed for IoT products such as smart lighting, smart locks, and security and alarm among many others.

This sub-1 GHz band technology is designed for reliable and low-latency communication of small data packets. It is scalable (supports up to 232 devices) and supports mesh networks without the need for a coordinator node.

5.4.4.2 Medium Range

5.4.4.2.1 Wi-Fi (Medium Range, High Data Rate, High Power) The same good old technology that connects most of our computers and gadgets to the Internet can be used to connect IoT and wearable devices as well. Because Wi-Fi consumes a relatively higher energy compared to other technologies, it is often overlooked for battery-operated devices, but its pervasiveness and low cost make it a viable option for certain applications. Wi-Fi, depending on the operating frequency (2.4 and 5 GHz bands) and number of antennas can offer different ranges (up to 70 m indoors), and data rates (600 Mbps maximum, but 150–200 Mbps is more typical).

5.4.4.2.2 ZigBee (Medium Range, Low Data Rates, Low Power) ZigBee-based networks are characterized by low-power consumption, low data rates (up to 250 kbps), and a line of sight connectivity range of up to 300 m, and 100 m for indoors.

The ZigBee standard is a relatively simple, easy to install, scalable to thousands of nodes, resistant to communication errors and unauthorized readings, and has high security and robustness. Typical applications include wireless sensor networks (WSNs) in M2M, IoT, and wearable technology applications.

5.4.4.3 Long Range

5.4.4.3.1 LPWAN and LoRa (Long Range, Low Data Rate, Low Power) Low-power wide-area network (LPWAN) is a type of network that serves the needs of applications requiring long-distance communications but also with limited power budget.

LoRa (Long Range) is an LPWAN technology that uses license-free subgigahertz bands like 433, 915, and 923 MHz. LoRa enables long range (2–5 km in urban areas, 15 km in suburban areas) with low-power consumption.

5.4.4.3.2 Sigfox (Long Range, Low Data Rate, Low Power) Sigfox is an ultra-narrowband (UNB) technology based on binary phase-shift keying (BPSK). Sigfox encodes the data by taking very narrow slices of spectrum and changes the phase of the RF carrier signal. This allows the receiver to only tune in to a small slice of the spectrum aiming at mitigating the effect of noise. Achievable data rate in

Sigfox is modest (up to 1 kbps) but it can support a wide range of up to 50 km in open areas with a very low-power consumption.

5.4.4.3.3 Cellular Technology (Long Range, High Data Rate, High Power) Cellular technology is the basis of mobile phone networks but it could also serve as a platform for IoT applications that require long-distance communication. Cellular technology is capable of transferring large amounts of data but at the expense of high-power consumption and cost.

Global System for Mobile Communications (GSM) has also been used for IoT systems represented by Extended Coverage GSM IoT (EC-GSM-IoT) which is a standard-based low-power wide area network technology. It is based on e-General Packet Radio Services (eGPRS) and implemented as a long-range, high capacity, and low energy cellular system for IoT communications. 4G Long Term Evolution (LTE) networks also support IoT. However, the exponential growth in IoT market has kept LTE networks struggling to keep up with the resource demands. The

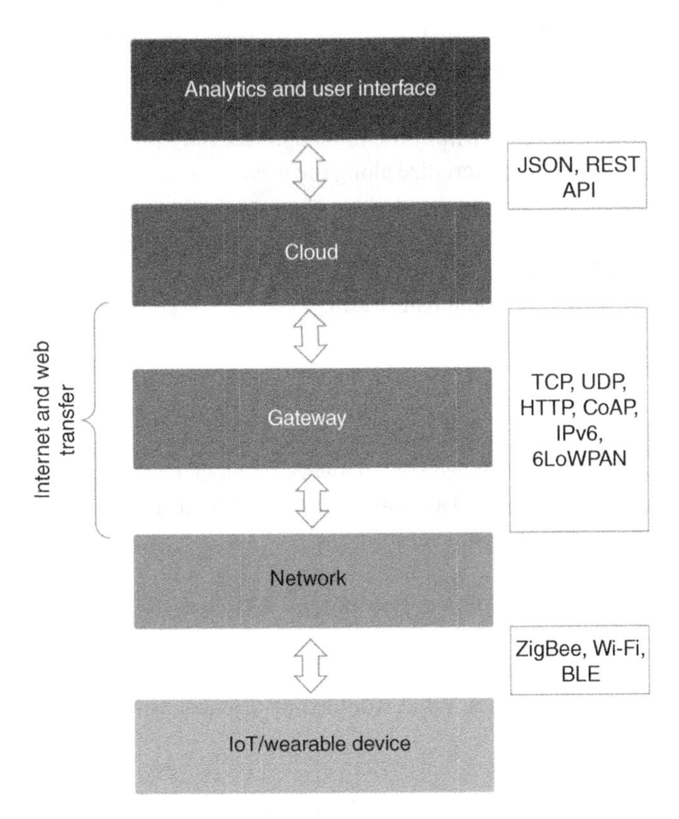

Figure 5.5 An example of an end-to-end IoT/wearable device connectivity.

solution is: 5G. As described by the International Telecommunication Union (ITU), all usage scenarios for 5G networks support IoT devices: massive Machine-Type Communications (mMTC), enhanced Mobile Broadband (eMBB), and Ultra-Reliable and Low-Latency Communications (URLLC).

Examples of cellular IoT include Cat-1 networks which are easy to set up and offer a great solution for voice- and browser-based applications and NB-IoT/Cat-M2 which uses Direct-Sequence Spread Spectrum (DSSS) modulation to send data directly to data centers without the need for a gateway. Although NB-IoT is not a cost-effective solution, eliminating the need for a gateway compensates for it.

Figure 5.5 shows a stack example of an end-to-end connection.

5.5 Conclusion

Today, IoT and wearable devices support a plethora of different protocols. In light of this, many technical bodies have started to call for a universal protocol standardization. However, being inherently scattered, the market of IoT and wearables will probably never be in real need of a unifying standard. Just as there are newer applications and use cases emerging within the industry, necessary protocols for their deployment will continue to materialize along the way.

On the other hand, selecting the appropriate type of connectivity is an inevitable part of any IoT or wearable technology project. It was demonstrated in this Chapter that the available connectivity options are governed by a trade-off between power consumption, range, and bandwidth.

Problems

1 You have narrowed down your choice for a network topology to either a full mesh topology or a star topology. Determine how your final decision will affect deployment cost and communication speed.

2 Based on a literature research, comment on how IP addresses are arranged and displayed.

3 You are designing a fitness tracker. What would your protocol and topology choices be?

4 Sketch a protocol stack for a smart IoT-based thermostat.

5 Sketch a protocol stack for a smart watch. Compare the flow of data with the one you sketched in the previous question.

6 Explain why it is better to use 6LoWPAN-UDP-CoAP stack in IoT instead of a stack of IPv6-TCP-HTTP.

7 A network with all the nodes acting as both servers and clients. A PC can access files located on another PC but also delivers files to other PCs on the network. Which network architecture is that?

8 Which of the following is NOT an advantage of a star network topology?

a) There is no central point of failure
b) Easy to add or remove a node as it has no effect on any other node
c) Reasonable security, i.e. no node can interact with another without passing through the server first
d) A few data collisions as each node has its own connection to the server

9 Which layer does the Ethernet and Wi-Fi protocols belong to?

10 What happens to the packet as it is passed from the application layer to the transport layer? What about from the transport layer to the network layer?

Technical Interview Questions

1 What is a three-way handshake in TCP?

2 What is the importance of the physical layer in the OSI model?

3 What is the difference between UDP and TCP?

4 What are the different layers in the OSI model?

5 What is a VLAN? Explain the VLAN trunking protocol.

6 What are the differences between Unicasting, Anycasting, Multicasting, and Broadcasting?

7 What are the differences between Hub, Switch and Router?

8 What are the most important topologies in computer networks?

9 What is the main difference between baseband and broadband transmission?

10 Draw a diagram of a network you have worked on, and explain it in detail.

11 What are the advantages and disadvantages of mesh topology.

12 What are the primarily used protocols in IoT?

13 What are two of IoT protocols based on REST architecture?

14 Name an important IoT protocol based on Publish – Subscribe scheme.

15 Explain various types of networks based on their sizes?

16 What are some transport layer protocols used in IoT, how do they work?

17 Explain Bluetooth Low Energy protocol for IoT and wearable technology?

18 How is wireless sensor network (WSN) applied in IoT?

19 Define ZigBee? Why is it important in IoT?

20 State the differences between the client-server model and the P2P model.

Further Reading

Aazam, M. and Huh, E.-N. (2014). Fog computing and smart gateway based communication for cloud of things. *Proceedings of the 2nd IEEE International Conference on Future Internet of Things and Cloud (FiCloud '14)* (August 2014). Barcelona, Spain, pp. 464–470.

Atzori, L., Iera, A., and Morabito, G. (2010). The internet of things: a survey. *Computer Networks* 54 (15): 2787–2805.

Baronti, P., Pillai, P., Chook, V.W.C. et al. (2007). Wireless sensor networks: a survey on the state of the art and the 802.15.4 and ZigBee standards. *Computer Communications* 30 (7): 1655–1695.

Bonomi, F., Milito, R., Natarajan, P., and Zhu, J. (2014). Fog computing: a platform for internet of things and analytics. In: *Big Data and Internet of Things: A Road Map for Smart Environments, Studies in Computational Intelligence book series SCI*, vol. 546 (eds. N. Bessis and C. Dobre), 169–186. Berlin, Germany: Springer.

Chang, K.-H. (2014). Bluetooth: a viable solution for IoT? (industry perspectives). *IEEE Wireless Communications* 21 (6): 6–7.

Colitti, W., Steenhaut, K., De Caro, N. et al. (2011). Evaluation of constrained application protocol for wireless sensor networks. *Proceedings of the 18th IEEE Workshop on Local and Metropolitan Area Networks (LANMAN '11)* (October 2011). Chapel Hill, NC: IEEE, pp. 1–6.

Culler, D. and Chakrabarti, S. (2011). 6lowpan: incorporating IEEE 802.15. 4 into the IP architecture, IPSO Alliance. White Paper, 2009.

Eisenhauer, M., Rosengren, P., and Antolin, P. (2009). A development platform for integrating wireless devices and sensors into ambient intelligence systems. *Proceedings of the 6th IEEE Annual Communications Society Conference on Sensor, Mesh and Ad Hoc Communications and Networks Workshops (SECON Workshops '09)* (June 2009). Rome, Italy: IEEE, pp. 1–3.

Gomez, C., Oller, J., and Paradells, J. (2012). Overview and evaluation of bluetooth low energy: an emerging low-power wireless technology. *Sensors* 12 (9): 11734–11753.

Gubbi, J., Buyya, R., Marusic, S., and Palaniswami, M. (2013). Internet of Things (IoT): a vision, architectural elements, and future directions. *Future Generation Computer Systems* 29 (7): 1645–1660.

Han, D.-M. and Lim, J.-H. (2010). Design and implementation of smart home energy management systems based on ZigBee. *IEEE Transactions on Consumer Electronics* 56 (3): 1417–1425.

Hanes, D. (2017). *IoT Fundamentals: Networking Technologies, Protocols, and Use Cases for the Internet of Things*. London, UK: Pearson Education.

https://azure.microsoft.com/en-us/overview/internet-of-things-iot/iot-technology-protocols/

Hui, J.W. and Culler, D.E. (2008). Extending IP to low-power, wireless personal area networks. *IEEE Internet Computing* 12 (4): 37–45.

Hunkeler, U., Truong, H.L., and Stanford-Clark, A. (2008). MQTT-S— a publish/subscribe protocol for wireless sensor networks. *Proceedings of the 3rd IEEE/Create-Net International Conference on Communication System Software and Middleware (COMSWARE '08)* (January 2008). Bangalore, India, pp. 791–798.

Katasonov, A., Kaykova, O., Khriyenko, O. et al. (2008). Smart semantic middleware for the internet of things. *Proceedings of the 5th International Conference on Informatics in Control, Automation and Robotics (ICINCO '08)* (May 2008). Funchal, Portugal, pp. 169–178.

Khan, W.Z., Xiang, Y., Aalsalem, M.Y., and Arshad, Q. (2013). Mobile phone sensing systems: a survey. *IEEE Communications Surveys & Tutorials* 15 (1): 402–427.

Lane, N.D., Miluzzo, E., Lu, H. et al. (2010). A survey of mobile phone sensing. *IEEE Communications Magazine* 48 (9): 140–150.

Liu, H., Bolic, M., Nayak, A., and Stojmenovic, I. (2008). Taxonomy and challenges of the integration of RFID and wireless sensor networks. *IEEE Network* 22 (6): 26–32.

Locke, D. (2010). MQ telemetry transport (MQTT) v3. 1 protocol specification, IBM developer Works Technical Library. http://www.ibm.com/developerworks/webservices/library/wsmqtt/index.html.

Mashal, I., Alsaryrah, O., Chung, T.-Y. et al. (2015). Choices for interaction with things on Internet and underlying issues. *Ad Hoc Networks* 28: 68–90.

Mitrokotsa, A. and Douligeris, C. (2009). Integrated RFID and sensor networks: architectures and applications. In: *RFID and Sensor Networks: Architectures, Protocols, Security and Integrations* (eds. Y. Zhang, L.T. Yang and J. Chen), 511–535. Boca Raton, FL: Auerbach Publications.

Noury, N., Herve, T., Rialle, V. et al. (2000). Monitoring behavior in home using a smart fall sensor and position sensors. *Proceedings of the 1st Annual International IEEE-EMBS Special Topic Conference on Microtechnologies in Medicine and Biology (MMB '00)* (October 2000). Lyon, France, pp. 607–610.

Pantelopoulos, A. and Bourbakis, N.G. (2010). A survey on wearable sensor-based systems for health monitoring and prognosis. *IEEE Transactions on Systems, Man and Cybernetics Part C: Applications and Reviews* 40 (1): 1–12.

Razzaque, M.A., Milojevic-Jevric, M., Palade, A., and Cla, S. (2016). Middleware for internet of things: a survey. *IEEE Internet of Things Journal* 3 (1): 70–95.

Schmidt, A. and Van Laerhoven, K. (2001). How to build smart appliances? *IEEE Personal Communications* 8 (4): 66–71.

Serrano, M., Barnaghi, P., Carrez, F. et al. (2015). *Internet of Things Semantic Interoperability: Research Challenges, Best Practices, Recommendations and Next Steps*. Oslo, Norway: European research cluster on the internet of things, IERC.

Shanmuga Sundaram, B. (2016). A quantitative analysis of 802.11ah wireless standard. *International Journal of Latest Research in Engineering and Technology* 2: 26–29.

Shelby, Z., Hartke, K., and Bormann, C. (2014). The Constrained Application Protocol (CoAP). *Tech. Rep., IETF document RFC 7252.*

Sheng, Z., Yang, S., Yu, Y. et al. (2013). A survey on the ietf protocol suite for the internet of things: standards, challenges, and opportunities. *IEEE Wireless Communications* 20 (6): 91–98.

Song, Z., Cardenas, A.A., and Masuoka, R. (2010). Semantic mid- dleware for the internet of things. *Proceedings of the 2nd International Internet of Things Conference (IoT '10)*, Tokyo Japan (December 2010).

Stanford-Clark, A. and Linh Truon, H. (2008). MQTT for sensor networks (MQTT-S) protocol specification, International Business Machines Corporation Version 1.

Sun, W., Choi, M., and Choi, S. (2013). Ieee 802.11 ah: a long range 802.11 WLN at sub 1 GHz. *Journal of ICT Standardization* 1 (1): 83–108.

Vasseur, J.P. and Dunkels, A. (2008). Ip for smart objects. White Paper 1, IPSO Alliance.

Vasseur, J.P., Bertrand, C.P., and Aboussouan, B. (2010). A survey of several low power link layers for IP smart objects. White Paper, IPSO Alliance. 24 Journal of Electrical and Computer Engineering.

Vasseur, J., Agarwal, N., Hui, J. et al. (2011). Rpl: the ip routing protocol designed for low power and lossy networks. Internet Protocol for Smart Objects (IPSO) Alliance 36.

Vermesan, O., Friess, P., Guillemin, P. et al. (2011a). Internet of things strategic research roadmap. In: *Internet of Things: Global Technological and Societal Trends*, vol. 1 (eds. O. Vermesan and P. Friess), 9–52. Aalborg, Denmark: River Publishers.

Vermesan, O., Friess, P., Guillemin, P. et al. (2011b). Internet of things strategic research agenda. In: *Internet of Things -Global Technological and Societal Trends* (eds. O. Vermesan and P. Friess). Aalborg, Denmark: River Publishers.

Villaverde, B.C., Pesch, D., De Paz Alberola, R. et al. (2012). Constrained application protocol for low power embedded networks: a survey. *Proceedings of the 6th International Conference on Innovative Mobile and Internet Services in Ubiquitous Computing (IMIS '12)* (July 2012). Palermo, Italy, pp. 702–707.

Weyrich, M. and Ebert, C. (2016). Reference architectures for the internet of things. *IEEE Software* 33 (1): 112–116.

Whitmore, A., Agarwal, A., and Da Xu, L. (2015). The internet of things—a survey of topics and trends. *Information Systems Frontiers* 17 (2): 261–274.

Wu, M., Lu, T.-J., Ling, F.-Y. et al. (2010). Research on the architecture of internet of things. *Proceedings of the 3rd International Conference on Advanced Computer Theory and Engineering (ICACTE '10)* (August 2010). Chengdu, China: IEEE, vol. 5, pp. V5-484–V5-487.

6

Product Development and Design Considerations

6.1 Introduction

The world of IoT and wearable technology is rapidly growing and steadily pushing for new innovative products. If these devices did not provide the potential of an immense value at a low cost, there would not be discussions about developing solutions based on these technologies in the first place. In fact, the demand is ongoing and the market is very exciting; however, product engineers and designers face new challenges and design constraints.

With more connected devices coming to the market every day, it is extremely important to ensure their functionality, security, and interoperability. Whether creating a new smart connected product or incorporating a new technology into existing products, there are key considerations to make. For example, wearables are generally characterized by portability, flexibility, and multi-functionality compared to handheld devices. Moreover, particular performance capabilities must be integrated into compact form factors. Therefore, the design process, materials selection, and manufacturing and packaging methods could be quite unconventional at times and need to be addressed and evaluated.

This chapter discusses the development process and design considerations that developers must follow to guarantee a successful launch of IoT and wearable products.

6.2 Product Development Process

When developing a new product, there is a set of steps that must be followed to turn an innovative idea into a product available for sale. Some steps can have multiple iterations, which is typical when it comes to developing technically complex products based on hardware and software systems.

Fundamentals of IoT and Wearable Technology Design, First Edition. Haider Raad.

Published 2021 by John Wiley & Sons, Inc.

6.2.1 Ideation and Research

The development process starts with identifying an idea for a new product, which could be either an enhanced version of an existing product, or a nonexistent product driven by a need. This step is typically followed by research and feasibility study which involves identifying the technology, materials, and methods to realize the end product.

6.2.2 Requirements/Specifications

The outcome of the previous step results in particular design specifications, a set of engineering requirements, along with an estimate of the cost of the end product. Section 6.3 of this chapter discusses the general product requirements in detail.

6.2.3 Engineering Analysis

Engineering analysis involves the application of scientific principles and analytical methods to understand and analyze the properties and mechanisms of a system. This is enabled by breaking the system down into basic components to understand their features and relationships with each other. In the development of modern electronic devices development, this step is generally divided into three sections:

- Hardware\Electrical design
- Software\Embedded System Design
- Mechanical Design

A large number of activities take place during this phase of the project. Many of them need to be carried out and coordinated in parallel. It should also be noted that when developing an electronic device, it is important to develop a test and production strategy alongside.[1] The realization and acquiring of the equipment needed in the design and development process may take place during this stage.

6.2.3.1 Hardware Design

This is often the main emphasis of the development process. It begins with the top-level design, and then the requirements are broken down into smaller subsections. At this stage an electrical schematic diagram is created, the layout for a

1 **Design for manufacture (DFM)** is the aspect of the design process where consideration is given to ensure ease of manufacturing processes aiming at minimizing the production cost.

printed circuit board (PCB) is designed, and the first draft of Bill of Materials (a list of components to be used in the product development) is generated. Controls, functionalities, and user interface design are all designed in this step.

6.2.3.2 PCB Design

Generally, the PCB realization is a major step of the electronics hardware development, and is often, the most time-consuming. Signal integrity analysis is also conducted as part of this step.[2]

IoT and wearable products come in all shapes and sizes, and to meet the form-factor and ergonomic requirements of specific applications, it becomes inevitable to use multiple PCBs.

Some of the drawbacks that come with using multiple PCBs are

- occupying additional space
- introducing additional point(s) of failure
- introducing additional assembly steps
- additional costs due to PCB connectors and cables

To overcome such drawbacks, rigid-flex PCBs are increasingly being used nowadays. These PCBs utilize a flexible polyimide layer embedded in the stack-up to hold interconnecting copper layers between the different sections allowing the finished assembly to be folded. Components are then mounted on the rigid sections in the traditional way, but the flexible sections can be buried internally within the rigid section.

Obviously, PCBs are designed using PCB CAD software packages and hardware tools. Also, many modern professional PCB CAD systems support 3D mechanical designs to be imported for a more realistic consideration. It is worth noting that if any RF component, such as the antenna feeding element, is incorporated in the PCB then the CAD system must support RF design parameters, in particular, impedance matching and return loss.

6.2.3.3 Software Development

Determining the operating system platform and the requirements for the device's software is a crucial step in the development process. Well-defined software specifications at this step will not only reduce the number of iterative test cycles, but

---

2 **Signal integrity** deals with the electrical performance of the wires, conductive tracing, and other structures used to carry signals within an electronic product. At high bit rates (high-frequency clock), various effects can degrade these signals to the point where errors take place, and the system could fail. Signal integrity engineering deals with analyzing and mitigating such effects. It is an essential task at all levels of electronics packaging and assembly.

will also provide a clear perspective of what the essence of the product is. This step should start with the top-level design, breaking down the requirements to smaller tasks that can be tackled separately.

One of the most important decisions to make before the start of product development is determining whether an operating system will be used. The choice of real-time operating system (RTOS), or high-level operating system (HLOS), development packages, programming languages, use of third-party libraries such as networking stacks, and GUIs must be determined as well. Such decisions will have an impact on selecting the microprocessor and memory of the product.

6.2.3.4 Mechanical Design

The mechanical design is also an important step in the overall device development process. It does not only deal with designing a mechanical enclosure or packaging, but important aspects such as heat flow and cooling analysis, force distribution, and mechanical interfaces are all evaluated in this step. This is usually done using mechanical modeling and simulation software packages.

It should also be noted that for those working in regulated industries, it is even more critical to use product lifecycle management (PLM)[3] to push requirements that comply with regulatory and safety standards.

Next, schematic diagrams of each block of the system of the electronic design are laid out. Once schematics are ready, the design for the actual PCB is created. The PCB serves as the physical platform that holds and connects all of the integrated circuits and electronic components. Moreover, code pieces are verified, and all product materials are determined in this step.

6.2.4 Prototyping

Prototyping generally refers to creating a sample model of a new product or process for testing and evaluation purposes. Prototyping provides specifications for a physical, working system rather than a virtual, theoretical one.

Some preliminary prototypes are basic and simple, and intended to visualize how a product might work, while others represent an actual representation of the end product. The selected order of a prototype, which is cost-dependent, must fit the specific requirements of the tests. The device enclosure, user interface designs, and the software deployment, which was defined in the previous step, are also implemented in this phase.

---

3 **Product life cycle management (PLM):** The development of IoT and wearable devices requires electrical, mechanical, and software design teams to collaborate together beginning from the earliest stages of the project. Product life cycle management (PLM) solutions are particularly designed to help bring all teams and designs together into a single system to enable faster design approvals and improved traceability from concept to final product launch.

6.2.5 Testing and Validation

Testing and validation involve evaluating the prototype to determine whether the product satisfies all the requirements and specifications defined in the second step. Testing is carried out at the very end of the hardware and software development process and is usually assigned a relatively smaller amount of time.[4]

Most testing and validation procedures encompass the following:

6.2.5.1 Review and Design Verification

At this stage, all the processes carried out previously are reviewed before starting the actual testing. This test is often called the Test Readiness Review (TRR).

Design verification ensures that the theoretical design meets the product requirements within the expected manufacturing and component tolerances. Specific areas of inspection typically include:

- Power supply check: Ensuring that voltages, currents, and power dissipations are correct. Supply rail sequencing and reset timings must also be checked.
- High-speed interfaces: Checking USB, Serial Advanced Technology Attachment (SATA), and memory interfaces.
- RF subsystem: Ensuring RF components are working in the design within their recommended operating conditions.

6.2.5.2 Unit Testing

Unit testing involves taking individual components and modules of the product, isolating them from the rest, and making sure they are functioning exactly as intended. This step is essential to ensure that each component of the main sections (software, hardware, and mechanical) meets its specifications to prevent issues during the final assembly.

6.2.5.3 Integration Testing

In this step, components and modules that have been tested individually are integrated and tested as a single unit. This step ensures that the interface and interaction between all parts of the device are appropriate and error-free. Life cycle testing may also take place at this point.[5]

---

4 **Regulatory Pre-compliance Testing**: The goal of this testing is to detect in the early stages if there are EMC or safety issues that need to be fixed. Preparation from the development team who typically support a variety of test modes is required for both pre-compliance and official compliance testing.

5 **Life cycle testing:** typically involves testing the product under operating conditions significantly beyond the norm. This type of testing in the design phase is commonly known as highly accelerated life testing (HALT), and when conducted on production samples is known as highly accelerated stress screening (HASS).

6.2.5.4 Certification and Documentation

Safety and compliance certifications, such as FCC, FDA, and CE, are generally required for new electronic products. These certifications provide verification that all the regulatory compliance requirements have been satisfied in order for a product to be distributed and sold legally. For example, in Canada and Europe, electronic products require both electromagnetic compatibility (EMC) and safety testing before they can be marked for sale, while this is regulated by the Federal Communications Commission (FCC) in the United States. At this stage of testing, the class of the device must be defined and appropriate testing organizations must be identified.

6.2.5.5 Production Review

A production readiness review should be conducted before the product is forwarded to production. This review marks the last step of the testing phase which is intended to ensure that the product development has been satisfactorily completed.

6.2.6 Production

The product can enter the production phase once a production readiness review is completed. The purpose of this stage is the industrial production of the device and making it available for purchase to the end user. Figure 6.1 depicts a general product development process diagram for modern electronic devices.

6.3 IoT and Wearable Product Requirements

When pursuing a new product development, it is essential to define the requirements which are typically captured from trial users, end customers, or market assessments.

The requirements are typically documented and serve as an agreement between the client and the product engineering team. The document can be used towards the product delivery time as a checklist for product completeness upon delivery.

While IoT and wearable technology could help create greater user experiences and better customer satisfaction, there also exists a need to handle the requirements that define how the capabilities of the design will work in the first place. On the other hand, in the rush to get innovative products to the overly competitive market, some key requirements can be overlooked, putting security and other design features at risk. Below are some major requirements the product design/development team should pay attention to:

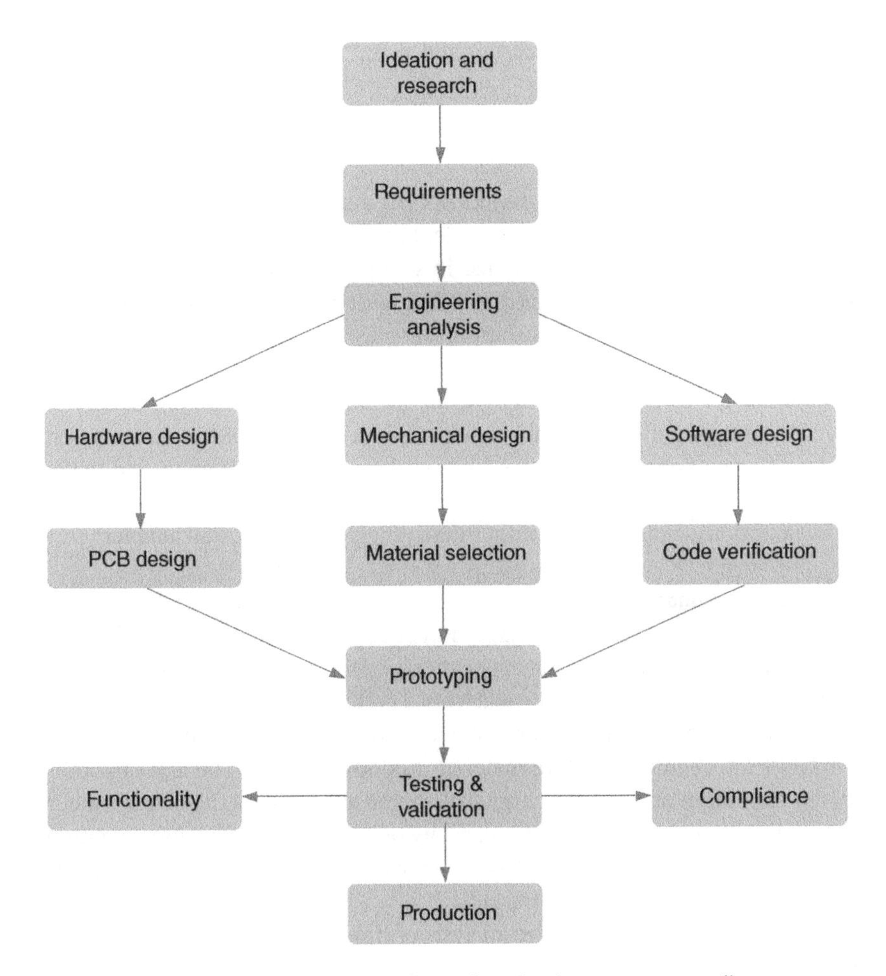

Figure 6.1 A general modern electronic product development process diagram.

6.3.1 Form Factor

Form factor is a hardware design aspect in electronics packaging which specifies the physical dimensions, shape, weight, and other components specifications of the PCB or the device itself.

Although wearables have a small form factor in general, it is truly dependent on the type and the way they are worn (i.e., rings and wristbands, versus glasses and jackets). This is also true for IoT devices, i.e., a compact smart metering device as opposed to a smart appliance or industrial equipment.

Devices with smaller form factors may offer reduced usage of material, easy handling and use, and typically low power consumption; however, they are typically associated with higher design and manufacturing costs in addition to maintenance constraints.

6.3.2 Power Requirements

Although some IoT and wearable devices operate autonomously such as the solar-powered trackers, the majority are dependent on batteries as an energy source.

The choice of battery type and size in compact portable IoT devices and light-weight wearables is vital and is strongly dependent on the operational needs (expected operational duration, display utilization, computational power, etc.).

Another factor to be considered is recharge-ability. Typically, the battery is charged by plugging the device into an adapter or a powered USB port. However, the demand for wireless charging has increased recently since it offers additional convenience. For instance, users find it easier to just drop their charge-needing smartwatches on a charging base rather than plugging it into a wall adapter.

6.3.2.1 Energy Budget

As mentioned earlier, wearables and hand-held IoT products are typically powered by a battery. IoT and wearable applications will be rendered useless, if battery life is unreliable and/or short. The capacity of a battery (typically in Ampere hour) is a measure of the amount of charge stored by the battery and is determined by the mass of the chemically active material inside the battery. The capacity indicates the maximum amount of energy that can be delivered by the battery under specified conditions. A battery with a capacity of 2000 mAh (milli Ampere hours) means that the battery can deliver 2000 mA current within one hour, 1000 mA for two hours, or 500 mA for four hours, etc.

The power budget deals with the analysis of how much power a given device requires for operation. Here, this analysis is required to determine how long a portable IoT or wearable device will operate from a battery of a given capacity before turning off. This is determined by calculating how much time a device will spend in each of its operating modes and then summing the energies expended in each mode.

Example A primitive wearable device operated by a CR2032 battery with a capacity of 0.225 Ah. The device consumes 1 ms operating time (on) for every two seconds with a run current I_{run} of 8.2 mA and sleep current I_{slp} of 1 μA. How long can the device be used before the battery has to be replaced?

Sleep time $t_{slp} = 1.999\,\text{s}$
Run time $t_{run} = 0.001\,\text{s}$
Sleep current $I_{slp} = 1\,\mu\text{A}$
Run current $I_{run} = 8.2\,\text{mA}$

$$\text{Average current } (I_{avg}) = \frac{I_{slp} * t_{slp} + I_{run} * t_{run}}{t_{slp} + t_{run}} = 5.1\,\mu\text{A}$$

CR2032 battery capacity $C = 0.225\,\text{Ah}$

Average device current I_{avg} = Battery capacity/operating time

$5.1 \times 10^{-6} = 0.225\,\text{Ah/Operating Time}$

$\rightarrow$Operating time = 44 117 hour

44 117 hour = approximately five years[6]

6.3.3 Wireless Connectivity Requirements

With most IoT and wearable products having one or more wireless interfaces, an important decision to make is whether or not the onboard wireless systems should utilize original equipment manufacturer (OEM) modules or if they should undergo a custom design. Another factor to consider is software support, i.e., modules may be provided with a certified protocol stack (such as BLE, cellular, and WiFi) that can significantly reduce the amount of overhead for software development and testing.

In wearable technology, there is a number of additional challenges that engineers face when designing antennas and wireless systems that do not exist in conventional wireless system design. For example, the degradation in the resonant frequency and return loss of wearable antennas need to be considered since they are prone to shift/deterioration due to impedance mismatch if the antenna unit is flexed or bent. Moreover, radiation patterns distortion and gain deterioration are also likely to occur. Another crucial constraint that needs to be accounted for is the close proximity of the antenna to the user's body, which implies two issues: degrading the impedance matching of the antenna due to the high water content (higher electrical conductivity) of the human tissues; and the increased amount of electromagnetic power deposition in the tissues, which gives rise to health hazards due to hyperthermia.

Designers must also ensure that no over-limit radiation of electromagnetic waves is taking place, which is characterized using a standard procedure known as Specific Absorption Rate (SAR) test. It is also worth mentioning that other factors need to be considered in some situations where the antenna must withstand higher temperature, pressure, and humidity.

6.3.3.1 RF Design and Antenna Matching

Having a high RF radiation efficiency is extremely important in battery-powered IoT and wearable products. Without efficiency optimization driven by the impedance matching, the antenna and its RF circuitry can waste significant amounts of

---

6 **Note:** In practice, the designer should pay attention to the battery self-discharge rate which is a phenomenon in batteries in which internal chemical reactions decrease the stored charge (capacity) of the battery even when not used in a circuit.

power. For wireless designs with antenna(s) mounted on the PCB, both the feeder(s) and the antenna(s) will require impedance matching to ensure maximum radiation efficiency and minimize signal reflection back to the transceiver.

The integration of the RF transceiver with other subsystems in close proximity within a small-form factor product poses a major challenge: electromagnetic interference. The negative impacts are summarized below:

- Interference due to the coupling of unwanted signals into the antenna and its feeding port, which compromises the range either as a result of reduced receiver sensitivity or lowered signal to noise ratio
- Electronic and thermal noise caused by the microcontroller/microprocessor, power supplies or other subsystems being coupled into RF system through their control interfaces.

The reader is referred to antenna design and RF circuits books for theory and design procedures.

6.3.3.2 Link Budget

Link budget is a commonly used metric to evaluate the range of a wireless system. All gains and losses from the transmitter to the receiver over the air-interface must be taken into consideration in order to calculate the link budget.

Link budget accounts for the attenuation of the transmitted signal due to propagation, cable and connector losses, radiation efficiency, in addition to gains from antenna topology, repeaters, and amplifiers. Effects of channel fading should also be taken into account and can be manipulated by using techniques such as antenna diversity and multiple input multiple output (MIMO), and frequency hopping.

The basic equation for a link budget is based on Friis equation, and given as:

$$\textbf{Received power}(\textbf{dB}) = \textbf{Transmitted power}(\textbf{dB}) + \textbf{Gains}(\textbf{dB}) - \textbf{Losses}(\textbf{dB})$$

First, one should start with the transmitted power at the source then add in the gain from antennas and repeaters. Next, the losses of the cables, connectors, and anything the transmitted signal passes through (channel) are subtracted.

Friis equation is used in telecommunications engineering, where the received power by the receiving antenna is calculated under idealized conditions due to a specific power transmitted by another antenna some distance away. Friis' transmission equation for free space propagation is given below:

$$P_r = P_t + G_t + G_r + 20\log\left(\frac{\lambda}{4\pi}\right) - 20\log D$$

where P_t is the transmitted power, P_r is the received power, G_t is the transmitting antenna gain, G_r is the receiving antenna gain, λ is the wavelength,[7] and D is the

---

7 **Wavelength** can be obtained from the frequency of the electromagnetic wave: $C = \lambda \times F$, where C is the speed of light.

distance between the transmitter and the receiver. For example, a link budget of 120 dB at 433 MHz gives a range of approximately 2 km.

It should be noted that the decibel (dB) scale is widely used in electronics, signal analysis, and communication systems. The dB is a logarithmic way of describing a ratio especially when the range is extremely wide. The ratio may be power, voltage, some intensity, etc.

When we convert a value V into decibel scale, we always divide by a reference value V_{ref}, thus the quantity is dimensionless since it represents a ratio:

$$\frac{V}{V_{\text{ref}}}$$

The value in dB is given as:

$$V(\text{in dB}) = 10 \log \frac{V}{V_{\text{ref}}}$$

Power is normally measured in Watt (W) and milliwatt (mW). The corresponding dB conversions are dBW and dBm. The reader should be familiar with such conversions when working in this area.

For example: Sensitivity level (the threshold of receiving a signal) of a GSM receiver is 6.3×10^{-14} W which is equivalent to −132 dBW or −102 dBm; Bluetooth transmitted power is 10 mW which is equivalent to −20 dBW or 10 dBm; GSM mobile transmitted power is 1 W which is equivalent to 0 dBW or 30 dBm, etc.

Figure 6.2 expresses a link budget elaboration between a transmitter and a receiver.

6.3.3.2.1 *Tips*

- A 6 dB improvement gives rise to doubling the range.
- Doubling the frequency gives rise to half the range.

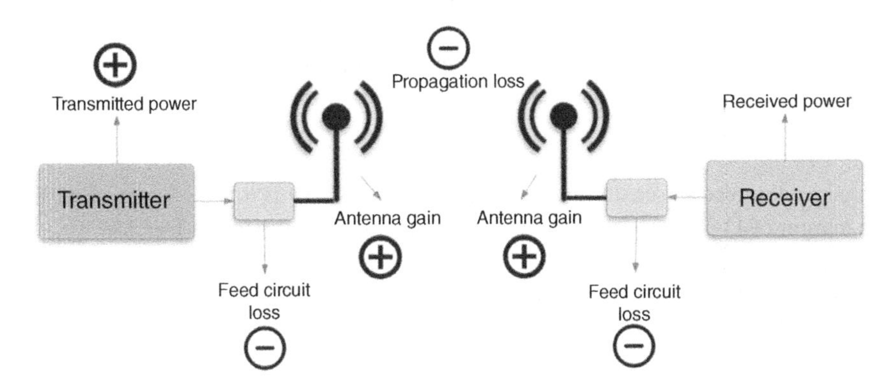

Figure 6.2 A link budget elaboration between a transmitter and a receiver.

Example A 2.4 GHz antenna of an access point has a gain of 10 dBi, a transmitting power of 20 dBm (equivalent to 100 mW), and a receiving sensitivity of −89 dBm. Five kilometers away, there's a stationary IoT device equipped with a 2.4 GHz antenna of 14 dBi gain, a transmitting power of 30 mW (15 dBm), and a receiving sensitivity of −82 dBm. The cables and connectors have a loss of 2 dB at each end. Is the communication link feasible?

Adding up all the gains and subtracting all the losses for the access point to the IoT device link gives:

$$\begin{aligned}&20\,\text{dBm}\left(\text{Transmit Power of access point}\right)+10\,\text{dBi}\left(\text{Access Point Antenna Gain}\right)\\&\quad-2\,\text{dB}\left(\text{Cable Losses for Access Point Antenna}\right)\\&\quad+14\,\text{dBi}\left(\text{IoT Antenna Gain}\right)-2\,\text{dB}\left(\text{Cable Losses IoT}\right)\\&=40\,\text{dB}=\text{Total Gain}\end{aligned}$$

The path loss for a 5 km link (λ is equal to 0.125 m at 2.4 GHz), considering only the free space loss is:

$$\text{Path Loss}=20\log\left(\frac{\lambda}{4\pi}\right)+20\log\left(5000\right)=113\,\text{dB}$$

Subtracting the path loss from the total gain

$$40-113\,\text{dB}=-73\,\text{dB}$$

Since −73 dB is greater than the minimum receive sensitivity of the IoT device (−82 dBm), the signal level is sufficient for the IoT antenna to be able to receive the access point's signal. There is only 9 dB of margin (82 − 73 dB) which will likely work fine in fair weather, but may not be enough during extreme weather conditions.[8]

Next, we calculate the link from the IoT device back to the access point:

$$\begin{aligned}&15\,\text{dBm}\left(\text{Transmit Power IoT device}\right)+14\,\text{dBi}\left(\text{Antenna Gain IoT device}\right)\\&\quad-2\,\text{dB}\left(\text{Cable Losses IoT Device}\right)+10\,\text{dBi}\left(\text{Antenna Gain Access Point}\right)\\&\quad-2\,\text{dB}\left(\text{Cable Losses Access Point}\right)=35\,\text{dB}=\text{Total Gain}\end{aligned}$$

Obviously, the path loss is the same in the opposite direction. Thus, the received signal level on the access point side is:

$$35-113\,\text{dB}=-78\,\text{dB}$$

Since the receive sensitivity of the access point is −89 dBm, this leaves us with a margin of 11 dB (89 − 78 dB). One could conclude that the link is feasible.

---

8 **Fading** is often modeled as a random process. A fading channel is a communication channel that experiences attenuation due to weather, obstacles, reflection, etc. Fading models based on statistical methods are available in the literature.

6.3.4 Cost Requirements

The cost of the product includes the initial outlay for the hardware and associated components (i.e., sensors, microcontrollers, etc.) as well as their ongoing operating costs, such as maintenance and replacement. Licensing fees for platforms, components, and device drivers should also be considered.

6.4 Design Considerations

Design considerations are factors that may affect the product or system requirements, design, or operational concept and should be part of the systems engineering process.

6.4.1 Operational Factors

Operational requirements deal with the device's essential capabilities and performance measures such as effectiveness, speed, accuracy, resolution, and consistency. Generally, the hardware and software design steps satisfy such factors.

6.4.2 Durability and Longevity

The device's durability depends heavily on the mechanical robustness of the packaging/enclosure material and the components quality of the internal circuitry.

The potential failure points of devices associated with structure and component deterioration should be identified and well documented. For feasibility purposes, this could be achieved by testing a sample until failure (typically after a device has successfully survived the target lifetime).

For example, in some applications, smart devices have to operate in harsh environments (extreme temperatures, pressure, operation under water, vibration, corrosion, humidity, etc.). In other applications, devices could be deployed in remote, hard to reach locations making maintenance and reconfiguration very costly. Hence, rugged designs must be considered to prevent any service interruption.

It is also worth noting in wearable applications that when operating on the human body, bending, flexing, and twisting become inevitable. Due to these effects, the performance of the internal components of the device could potentially deteriorate. Hence, some tests are necessary to ensure operative reliability. For example, flexibility tests are conducted by repeated trials of the prototype under bending, stretching, and twisting to monitor for any deformations or discontinuities, and to ensure there are no wrinkles or permanent folds introduced which might compromise the functionality and aesthetics of the device.

6.4.3 Reliability

For a smart device to be successful, it must be precise, consistent, punctual, and reliable. This is essential for the user to establish confidence and trust in the device. Any error tolerances must be identified by the manufacturer before the release of the product and must be clearly disclosed to the user. Moreover, all the system components should be accurately integrated and field-tested to ensure a reliable performance. For example, the problem of electromagnetic interference (EMI) typically arises when multiple components are integrated in compact form factors which could negatively impact the device's performance.

EMI can also affect the accuracy of data acquisition and measurement in addition to the reliability of the communication signals. These risks can be eliminated by embedding all components within a specially designed low-EMI enclosure.

6.4.4 Usability and User Interface

Usability, within the context of consumer electronics, is often defined as the ease of use, handling, and learnability of an electronic device. When designing IoT and wearable devices, it is imperative for the device to swiftly deliver the task requested by the user. Moreover, the ability to easily navigate through the device's user interface strongly influences the user's engagement and interaction.

The majority of people think of user interfaces as just software or apps on smartphones. In reality, a user interface could be anything from voice and gesture control, switches and buttons, to touch screens and control panels. Designing a highly responsive, user-friendly device ensures higher adoptability rates and pleasure of experience.

6.4.5 Aesthetics

Users' tastes vary based on psychological, societal, and cultural perceptions. Users also differ physically in built and complexion. To establish an emotional connection between the wide spectrum of users and the product, wearables and applicable IoT devices with different styles and fashion personalities may need to be offered to the consumers.

6.4.6 Compatibility

As mentioned in the previous chapters, IoT and wearable devices are, in many cases, synced to a gateway (i.e., a smartwatch connected to a smartphone or tablet) for data processing and forwarding. The connection is typically carried out via

WiFi, Bluetooth Low Energy (BLE) or Near Field Communication (NFC). Sometimes even when products operate under the same protocol but use different versions can cause interoperability issues. Hence, these devices must be compatible with different operating systems that could be encountered when a connection is established (i.e., Android versus iOS).

6.4.7 Comfort and Ergonomic Factors

This is one of the most important design aspects especially in wearables and portable IoT devices. The device's weight, shape, size, and texture must be carefully considered.

Devices should fit users comfortably enabling usage and movement without any constraints.[9] For example, stretch-ability, temperature and breathe-ability balance in textile-based wearables play a vital role in the commercial success of such devices.

6.4.8 Safety Factors

Wearables and IoT devices are meant to enhance the quality of users' lives and must be designed with specific product safety requirements in mind. Physical and psychological harm due to misuse and/or abuse of the device, in addition to potential operation, manufacturing, and assembly hazards must be carefully evaluated. Examples include tests for device overheating, accidental electric shocks, excessive electromagnetic radiation, and material toxicity. It should be noted that the long-term physiological effects of these devices are not established yet and need to be properly researched. The psychological and social impacts of wearables will be covered in Chapter 8 of this book.

6.4.9 Washing Factors (Wash-ability)

Wearables that are based on fabrics are usually exposed to dirt, dust, and sweat, which might compromise their performance. Obviously, the operation and performance of wearables are required to be consistent after it is washed or soaked in water.

---

9 **Ergonomics** is a relatively new discipline that deals with designing products, systems, or processes with an eye towards ensuring a proper, comfortable, and convenient handling and interaction between the device and the user.A proper ergonomic design requires researching other disciplines such as anthropometry, which deals with studying body sizes and shapes of a population, biomechanics, environmental physics, and applied psychology.

6.4.10 Maintenance Factors

The expected shelf life of the product is often determined in the design phase. It is also determined if the product will be maintenance-free, or would require routine maintenance, and whether the maintenance is to be performed by the end user or by a professional technician.

When the product is released to the market, support provided by the releasing company is essential to evaluate and resolve any errors or bugs detected in the product. This process is also vital for improving the quality and usability of the product.

6.4.11 Packaging and Material Factors

This deals with the selection and assessment of the material types used in IoT and wearable devices (plastic, rubber, metal, fabric, wood, etc.). The material of the device enclosure, for example, must be strong enough to protect the internal electronic circuitry. Hence, developers must consider the mechanical factors that affect the end product quality such as, the tensile strength, density, rigidity, durability, hardness, flexibility, and stretch-ability. Further, wearable devices are typically in direct contact with the user; hence, the effects of the selected materials on the device usability such as the toxicity and biocompatibility issues must be evaluated (i.e., skin-related allergies and irritations).

6.4.12 Security Factors

As mentioned previously, IoT and wearable devices collect and share data across a variety of systems and platforms. Protecting these connected devices requires an understanding of the risks and impacts of cyber-attacks, awareness of vulnerabilities and the placing of plans to mitigate such scenarios without compromising the design.

Moreover, functionality in many cases presents a trade-off over security. Realizing a product that is secure but not practically usable can be as problematic as a product that is less secure but practical. Software, hardware, and information security must be planned carefully before a product is setup.

Related security requirements include:

- Ensuring the product has enough memory and computational power to be able to encrypt and decrypt data at the rate they are transmitted and received
- Ensuring the libraries of the software development support the required authorization and access control mechanisms

- Adopting standard devices that implement management protocols for securely registering new devices as they are added to a network to avoid spoofing,[10] and procure security updates

6.4.13 Technology Obsolescence

Ensuring the availability of all components used in the bill of materials when mass production begins is of paramount importance. It is also important to estimate the availability of components during the lifetime of the product to ensure continuity. Another important consideration is the lifetime of the component itself; as a rule of thumb, it should match or exceed the expected product lifetime.

6.5 Conclusion

In a competitive industry, the development of a new product may involve risks but also creates business opportunities. The stages of product development may seem like a long process, but they are introduced to save time and resources. A careful planning for a new product development processes along with testing and validation of prototypes are essential steps toward ensuring that the new product will meet the target market needs. This chapter discussed key engineering requirements and considerations for designing and deploying successful IoT and wearable products.

Problems

1 A smart thermostat system uses a temperature sensor and a microcontroller with a Bluetooth connectivity. What is the capacity of the battery that you would choose for the device to last at least six years? Assume that the processor clock frequency is 50 MHz, communication current is 3.5 mA, data logging current is 25 μA, sleep mode current is 2 μA, and wakeup time is 140 μs. Communications run for 0.25 s every hour, data logging runs for 20 ms every second.

2 A battery-operated IoT device must run for two years without replacing the battery, at I_{run} = 25 mA, with 1 ms operation for every two seconds and sleep current I_{slp} = 1 μA. Determine the required battery capacity?

---

10 **Spoofing** is the act of impersonating another device or user on a network in order to gain illegitimate advantage (i.e., steal data, inject malware, or bypass access controls).

3 Pick a wearable device of your choice then list three battery candidates available commercially for a reasonable operating time. Justify your battery choice according to the energy budget of the wearable device.

4 What is the wavelength at 900 MHz, 2.45 GHz, and 60 GHz? What is the path loss over 1 m, 100 m, and 1 km for these frequencies?

5 What is the maximum range of a fitness tracker connected to a smartphone using BLE. Explain in terms of link budget analysis.

6 A WiMax base station transmits at power levels of 43 dBm, with an antenna gain of 14 dBi, and a receiving sensitivity of −92 dBm. An IoT irrigation system is located two miles away with a dipole antenna of 1.76 dBi gain, a transmitting power of 16 dBm, and a receiving sensitivity of −88 dBm. The cables and connectors have a loss of 3 dB at each end. Is the communication link feasible?

7 Sketch a flow diagram for the development process of a basic fitness tracker.

8 List all the possible design considerations you believe they are appropriate for a smart T-shirt that measures heart rate, breathing rate, and temperature.

9 Sketch a flow diagram for the development process of an IoT-based security system.

10 List all the possible design considerations you believe appropriate for an automatic dog feeder. The device lets you feed your dog remotely, schedule meals, and control the portion size.

Interview Questions

1 What is the relationship between dBm, dBW, and Watt?

2 What are some of the basic checks that need to be made when laying out a microcontroller-based design?

3 How's the antenna gain related to path loss/wireless coverage?

4 What would you do to improve the link budget?

5 What are the components of a battery management system?

6 Describe a flow of a complete PCB design.

7 You are given a physical product (i.e., a fitness tracker). What do you like about the design, and what you dislike? How would you improve it?

8 What design considerations would you list for designing a smart baby crib that detects if the baby is fussing, choking, having a fever, or a wet diaper?

9 Draw an RF/antenna circuit schematic and describe how you would lay it out on a PCB.

10 What are the main steps in the UI design process?

Further Reading

Arora, S., Yttri, J., and Nilsen, W. (2014). Privacy and security in mobile health (mHealth) research. *Alcohol Research: Current Reviews* 36 (1): 143–151.

Balakrishnan, A. (1998). Concurrent engineering: models and metrics. Master dissertation. McGill University, Canada.

Belliveau, P., Griffin, A., Somermeyer, S., and Meltzer, R. (2002). *The PDMA Toolbook for New Product Development*. New York: John Wiley & Sons.

Campbell, J.L., Rustad, L.E., Porter, J.H. et al. (2013). Quantity is nothing without quality: automated QA/QC for streaming environmental sensor data. *BioScience* 63 (7): 574–585.

Cooper, R. (2001). *Winning at New Products: Accelerating the Process from Idea to Launch*, 3e. Cambridge, MA: Perseus Publishing.

Cooper, R. and Edgett, S. (2008). Maximizing productivity in product innovation. *Research Technology Management* 51 (2): 47–58.

Damm, O. and Wrede, B. (2014). Communicating emotions: a model for natural emotions in HRI. *HAI '14 Proceedings of the Second International Conference on Human-Agent Interaction*, Tsukuba Japan, pp. 269–272.

Khaleel, H.R. (2014). *Innovation in Wearable and Flexible Antennas*. Southampton, UK: WIT Press.

Lallemand, C. (2011). Toward a closer integration of usability in software development: a study of usability inputs in a model-driven engineering process. *EICS '11 Proceedings of the 3rd ACM SIGCHI Symposium on Engineering Interactive Computing Systems*, Pisa, Italy, pp. 299–302.

Lilien, G., Morrison, P., Searls, K. et al. (2002). Performance assessment of the lead user idea generation process for NPD. *Management Science* 8 (8): 1042–1059.

Marshall, R. (2019). IoT hardware from prototype to production, a guide to launching hardware based IoT products for startups and scaleups, Xitex Ltd Lawrence Archard Product Development, uPBeat, Steve Hodges.

Morabad, A.D. (2018). Key considerations for successful new product development. HCL White Paper.

Patel, S., Park, H., Bonato, P. et al. (2012). A review of wearable sensors and systems with application in rehabilitation. *Journal of Neuro Engineering and Rehabilitation* 9: 21.

Scarpino, M. (2014). *Designing Circuit Boards with EAGLE: Make High-Quality PCBs at Low Cost*, 1e. Upper Saddle River, NJ: Prentice Hall.

Sullivan, S. (2017). *Designing for Wearables: Effective UX for Current and Future Devices (Book)*. Sebastopol, CA: O'Reilly Media.

Ulrich, K.T. and Eppinger, S.D. (2011). *Product Design and Development*. New York: McGrawHill.

Vogel, D. (2010). *Medical Device Software Verification, Validation and Compliance (Book)*. Boston, MA: Artech House.

Weinger, M.B., Wiklund, M.E., and Gardner-Bonneau, D.J. (2010). *Handbook of Human Factors in Medical Device Design*. Cleveland, OH: CRC Press.

Wiklund, M.E., Kendler, J., and Strochlic, A.Y. (2015). *Usability Testing of Medical Devices*, 2e. Cleveland, OH: CRC Press.

7

Cloud and Edge: Architectures, Topologies, and Platforms

7.1 Introduction

After your IoT or wearable technology project is up and running, devices will start to generate vast amounts of data. An efficient, scalable, and cost-effective means will be needed for managing those devices and handling all that information and deliver the desired outcomes for you. When it comes to long-term storage, processing, and data analysis, nothing can beat the cloud.

By minimizing the need for on-premises infrastructure, the cloud has enabled businesses to go beyond the conventional applications of IoT and wearable devices and accelerated the large-scale deployment of these technologies. Moreover, as data from the physical world comes in various formats, cloud platforms offer a wide range of management solutions from unstructured bits of data, such as images or videos, to structured entities, and high-performance databases for telemetry data.

On the other hand, Edge computing where data are processed closer to the endpoints is increasingly being employed in IoT and wearable technology in order to cut down the latency and expedite the decision making process. Current deployments often employ a mix of cloud and Edge computing to get the best of the two worlds.

For example, health monitors and other healthcare wearable devices can save lives by instantaneously alerting medical staff when help is needed. Moreover, smart surgical assistive devices must be able to analyze data swiftly, safely, and accurately. If these devices strictly rely on transmitting data to the cloud for decision making, the results could be disastrous.

This Chapter provides an overview of cloud topologies and platforms, and an architectural synopsis of OpenStack cloud. Next, Edge topologies and computing technologies will be presented. It will be shown that the maximum value from an

Fundamentals of IoT and Wearable Technology Design, First Edition. Haider Raad.

Published 2021 by John Wiley & Sons, Inc.

IoT or wearable technology project can only be gained from an optimal combination of cloud and edge computing, and not by a cloud-only architecture.

7.2 Cloud

7.2.1 Why Cloud?

IoT and wearable technology cloud comprises the services and standards necessary for connecting, managing, and securing a wide spectrum of devices and applications enabled by these technologies, in addition to the underlying infrastructure required for processing and storing the data produced by these devices. The cloud enables businesses to leverage the potential of these technologies without having to build the necessary infrastructure and services from the ground up.

The cloud offers a more efficient, scalable, and flexible model for bringing the infrastructure and services to power IoT and wearable devices and their applications. Most of IoT is virtually limitless in scale, unlike most organizations' resources. The cloud computing model effectively takes in the ever-expanding scale of IoT and wearable devices, and it can do so in a cost-effective manner.

7.2.2 Types of Cloud

A number of cloud models, types, and services have evolved over the years to offer organizations the right solution for the right needs. Cloud computing is usually classified on the basis of model or on the service being offered.

Cloud models can be categorized into three major types: public, private, and hybrid. Sometime a fourth category appears in the literature, that is, community cloud. Depending on the type of data an organization is dealing with, they will want to compare these models in terms of the different levels of security and management required.

7.2.2.1 Private Cloud

A private cloud is one in which an organization has exclusive access to its infrastructure resources. Clients can choose to have the private cloud located on-premises or hosted by a third-party service provider.

The main advantage a private cloud model is that it provides greater security and assurance compared to a public cloud model since it is guaranteed that information is confined strictly to systems managed by the client. A major disadvantage of this model is that it can be expensive to install. Additionally, organizations are restricted to cloud infrastructure resources as specified in a legal agreement. The strict security of a private cloud can make it harder to access remotely too.

7.2.2.2 Public Cloud

Public clouds provide the infrastructure and services over the Internet and are hosted at the vendor's premises. The client has limited visibility and control over where the service is hosted. However, services can be used, and access is granted anytime and anywhere, as needed.

In addition to delivering services over the web, the public cloud model offers several other advantages including low cost of ownership, automated deployments, scalability, and reliability. Also, organizations do not have to worry about installation and maintenance.

One drawback of a public cloud, however, is that it is not the most viable model in terms of security, and often cannot meet some security regulatory compliance requirements. This is due to the fact that servers are dispersed geographically across multiple countries with various security regulations. Moreover, while this model is generally cost-effective, expenses can rise exponentially for large-scale usage. Amazon, Google, IBM, and Microsoft are among the major players in the public cloud realm.

7.2.2.3 Hybrid Cloud

The hybrid model uses both private and public clouds, held together by technologies that enable them to share data and applications. For example, the most important applications can be hosted on a private cloud to keep them more secure and to allow greater flexibility and more deployment options, while secondary data applications can be hosted on public clouds.

7.2.2.4 Community Cloud

A community cloud is collaborative endeavor in which infrastructure is shared between a number of organizations that have common goals, interests, and concerns. The shared infrastructure could be managed internally or by a third party.

7.2.3 Cloud Services

Based on the types of resources that are accessed as services, clouds are associated with different delivery models. Each model introduces additional services. These offerings are the value-add of cloud technology. These services should at least offset the capital expense an organization has to deal with when purchasing and maintaining such servers and data equipment. Cloud service models are categorized as follows:

7.2.3.1 Infrastructure as a Service (IaaS)

IaaS offers the most flexibility and scalability in the deployment of storage and virtualized computing resources toward supporting custom solutions.

Organizations may choose this model in order to benefit from lower prices, the ability to cluster resources, in addition to customized security. The most prominent examples of IaaS service are Amazon Web Services (AWS) and Microsoft Azure.

7.2.3.2 Software as a Service (SaaS)

SaaS is a method for delivering software applications and services over the web normally just by logging in and is generally charged on a subscription basis or as a pay per use. SaaS providers host and manage the software application and associated infrastructure and handle related maintenance such as software and security updates.

7.2.3.3 Platform as a Service (PaaS)

PaaS refers to cloud services that provide an on-demand environment for securely developing, testing, deploying, and managing software applications. PaaS provides developers with everything they need without having to worry about the provisioning of the underlying infrastructure. Services could be one or a mix of middleware, database management, analytics, operating system, etc. Examples of PaaS providers include Google App Engine, Oracle Cloud Platform, and Heroku.

7.2.3.4 Functions as a Service (FaaS)

FaaS, sometimes referred to it as server-less architecture, overlaps with PaaS by adding another layer of abstraction and focusing on building app functionality without spending time managing the infrastructure, setup, and other logistics. FaaS applications consume no IaaS resources until a specific function or event takes place.

7.2.4 OpenStack Architecture

OpenStack is an open-source framework for building and managing cloud computing platforms for public and private clouds. OpenStack is managed by the OpenStack Foundation and backed by some of the biggest players in the software development and web hosting industry. OpenStack supports the deployments of both private and public clouds. It is characterized by flexibility, scalability, simplicity of implementation, and high configurability.

As shown in the previous section, cloud computing can refer to different service models, OpenStack falls into the Infrastructure as a Service (IaaS) category. Providing infrastructure means that OpenStack makes it easy for clients to swiftly deploy new instances.

7.2.4.1 Components of OpenStack

OpenStack is made up of many different components. Because it is an open-source framework, anyone can add additional components to OpenStack based on a given need. However, nine key components have been identified by the OpenStack community. These components are part of the OpenStack's core and are "pre-packaged" within any OpenStack system. They are also officially maintained by the OpenStack community.

It is worth noting that all communications within the OpenStack components are performed through Advanced Message Queueing Protocol (AMQP) message queues.

- **Nova** is the main computing engine behind OpenStack. It is used to identify computation resources based on demand, and for deploying and managing the virtual machines and other instances to handle computing tasks.
- **Swift** is a widely used object storage system designed to store files, backups, photographs and videos, analytics, web content, etc. Rather than referring to files by their location on a disk drive, clients can instead refer to a unique identifier pertaining to the file or data cluster and let OpenStack decide where to store it. This promotes scalability since developers do not have to worry about the capacity on systems behind the software.
- **Glance** provides discovery, registration, and delivery services for disk and server images (virtual copies). The ability to immediately capture a server image and store it away is a powerful feature of the OpenStack cloud. Stored images can be used as a template to quickly deploy new servers. Glance can also be used to store and classify an unlimited number of backups. Glance interacts with Swift (the object store) to retrieve or store the images, while the Glance application program interface (API)[1] provides an interface for querying information about these images and allows clients to stream the images to new servers.
- **Cinder**, the block storage component in the OpenStack, acts as a storage and as a service for scenarios involving databases. It dynamically expands file systems, such as data lakes, which are of paramount importance in IoT and wearable technology applications.
- **Neutron** provides the networking capability for OpenStack. Its main objective is to ensure that each OpenStack component can intercommunicate quickly and efficiently. The entire network is configurable and provides services such as

---

1 **Application Programming Interface (API):** An API is a set of routines, protocols, and tools for creating software applications. In IoT and wearable devices, an API lets the user access the status, configure, and control the functionality of a device, such as tank level or heart rate readings. Of special relevance is API REST. Representational state transfer (REST) is a software architectural approach that defines a set of criteria to be used for creating Web services. REST based Web services, aka RESTful, provide interoperability between internet devices. RESTful API enables an interface through HTTP calls, to retrieve data or indicate the execution of some operation on the data.

Domain Name Services (DNS), Dynamic Host Configuration Protocol (DHCP), management of VLANs, firewalls, and gateways, and many other services.

- **Keystone** provides identity management services for OpenStack which is responsible for establishing user credentials, API client authentication, service discovery, and high-level authorization. It provides multiple means of access and maintains a central directory of users and their access privileges. Keystone supports Lightweight Directory Access Protocol (LDAP), Open Authorization (OAuth), OpenID Connect, and Structured Query Language (SQL).
- **Ceilometer** provides telemetry services, which enable data gathering, resource metering, and billing solutions to individual users of the cloud. It also maintains certifiable tracking of usage of each component of an OpenStack deployment.
- **Heat** provides the coordination of infrastructure resources to OpenStack deployments. It allows clients to store the requirements of a cloud application in a text file that dictates what resources are needed for a given application. Heat also provides an auto-scaling service that can be integrated with the OpenStack telemetry services.
- **Horizon** is the web-based dashboard for OpenStack. It allows the clients to view the various components of OpenStack discussed in this section. Clients can access all of the components of OpenStack individually through an API; moreover and however, Horizon provides an alternative user interface to show what happens in the cloud and to manage it as necessary. Horizon allows a third party to customize their GUI and to add their interactive tools and widgets to the dashboard. Most IoT and wearable applications that use cloud deployments will integrate some form of a dashboard with a set of features.

7.3 Edge and Fog

As we saw earlier, clouds are located at the last station of the data train originating from the IoT or wearable device, and sits over the WANs.

It should be noted that most wearables and some IoT devices utilize non-IP-based protocols such as ZigBee and BLE when operating within PANs, otherwise, the data travel through IP-based protocols on its way to the cloud. This is where the Edge gateway comes into play. It acts as a coordinator and translator between the two.

The above is not the only reason an Edge layer is needed. Latency and response time are crucial effects that triggered the need for Edge. As we saw in the previous chapters, latency of a millisecond can have disastrous effects in some applications, and that the cloud component introduces extra latency over the WAN.

Fog computing draws a parallel from the success of Hadoop and MapReduce, which are open-source software utilities that facilitate the operation of networks of too many nodes to solve problems involving significant amounts of data and

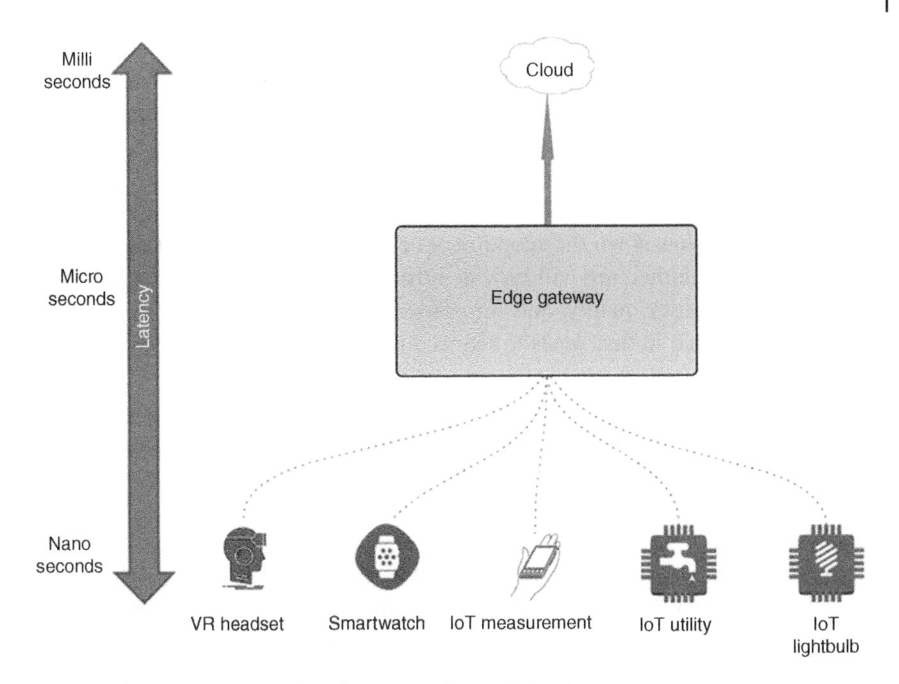

Figure 7.1 Latency comparison between edge and cloud.

computation. Hadoop is an open-source framework while MapReduce is a method of mapping. Hadoop is based on the MapReduce algorithm.

MapReduce is based on three steps: map, shuffle, and reduce. The map process applies computing functions to local data. During the shuffle phase, data are redistributed as needed. The reduced process applies processing across all the nodes in parallel.

In summary, MapReduce works to bring processing to where the data and not vice versa. This scheme effectively eliminates communication overhead and bottlenecks in systems that have massive amounts of structured or unstructured datasets (big data) which is also the case in the realm of IoT and wearables (Figure 7.1).

7.3.1 The OpenFog Reference Architecture[2]

The OpenFog Consortium, established by the industry's major players such as ARM, Cisco, Dell, Intel, and Microsoft, published the OpenFog Reference Architecture, a universal framework designed to fulfill the demanding

2 In-depth details of the full OpenFog reference architecture are published in a 160-page document [1].

requirements of smart devices, 5G, and artificial intelligence (AI) applications. As mentioned previously, a system-level architecture brings computing, storage, control, and networking functions closer to the data-generating sources as needed. The objective of such system-level architecture is to enable high-performance, interoperability, and security for the applications in reference.

The OpenFog Consortium was founded on the principle that proprietary vendor solutions would slow down the adoption of new technologies and further innovations. An open architecture will provide a robust platform for product development, promote better quality, and innovation through competition in an open environment which in turn leads to reduced product costs and improved market adoption.

The OpenFog Reference Architecture consists of layers extending from sensors and actuators at the bottom, to application services on top. The architecture has some common attributes with cloud architecture such as the OpenStack, and contains a medium to high-level view of system architectures for fog nodes (IoT and wearable devices) and networks, and deployment and hierarchy models. OpenFog is based on eight core technical attributes, referred to as "pillars". These pillars are as follows: security, scalability, openness, autonomy, reliability–availability–serviceability (RAS), agility, hierarchy, and programmability.

The OpenFog architecture provides a generic Fog platform that can be applied to any application and market. It aims at adding business value for IoT and wearable applications that are bound by network constraints, and providing real-time decision making, low data propagation latency, and improved security.

The OpenFog provides a full stack that comprises the following layers:

- **Application Services:** This layer acts as a connector to other services, a host for data analytics packages, a user interface if needed, and as a provider of core services.
- **Application Support:** This layer defines the components of the infrastructure required to build the final customer solution. Forms of support may include: runtime engines, login tools, application and web servers, application management, security services, and databases.
- **Node Management and Software Backplane:** This is also known as In-Band (IB) management which oversees how a fog node communicates with other nodes in the same domain. Through this interface, nodes are managed for upgrades, status check, and deployment. The software backplane can include the node's operating system along with necessary drivers and firmware, communication protocols and file system control.
- **Hardware Virtualization:** Just as in typical cloud systems, OpenFog reference architecture deals with the hardware as a virtualization layer. Applications should not be only compatible with specific sets of hardware.

- **OpenFog Node Security:** This layer defines the hardware security portion of the stack. In a given topology, upper level fog nodes should be able to oversee lower-level nodes as part of the hierarchy. Peer nodes should be able to monitor their neighboring nodes to the right and left. Encryption, physical tampering monitoring, and packet inspection are also among the responsibilities of this layer.
- **Network:** This is the first slice of the hardware layer. The network layer is aware of the Fog topology and routing, and has the role of physical routing to other fog nodes. This is a major distinction from typical cloud networks which virtualize all their internal interfaces. Here, the geographical location of the network has an impact on the performance of an IoT deployment.
- **Accelerators:** Another aspect of OpenFog that does not exist in cloud architectures is the use of accelerators such as general-purpose graphics processing units (GPGPUs) and field-programmable gate arrays (FPGAs) to provide services for imaging, digital signal processing, machine learning, and encryption and decryption.
- **Compute:** The compute slice of the stack is similar to the compute functionality in OpenStack. The key functions of this layer include task execution, resource provisioning, and load balancing.
- **Storage:** The storage portion of the architecture supports the low-level interface to the fog storage. This layer is also responsible of managing all the traditional types of storage devices, such as disk drives and RAM arrays.
- **Hardware Platform Infrastructure:** This layer does not deal with the actual software or hardware of the fog node but more with the physical structure and mechanical support. For example, fog devices could be installed in harsh and remote areas, and hence, they must be rugged and mechanically robust.
- **Protocol Abstraction:** The protocol abstraction layer bonds the bottom elements of the system (sensors or actuators) with other layers of the fog node and the cloud. By abstracting the interface between the layers, a heterogeneous combination of sensors can be deployed within a single fog node.
- **Sensors, Actuators, and Control Systems:** This is the bottom layer of the stack where the physical objects are laid. All of these objects (sensors, actuators, control elements, etc.) communicate with the fog node which has the responsibility to service, secure, and manage that device.

7.3.2 Fog Topologies

Fog topologies can exist in several forms, and a number of aspects such as cost, processing load, manufacturer interface need to be considered when designing an end-to-end fog system. A fog network can be as simple as an edge router connecting sensors or actuators to a cloud service, or as complex as a multi-tier fog hierarchy with different levels of computation capability at each tier simultaneously. Modeling factors are determined based on the following:

- **Data Reduction:** For example, is the smart system solution tasked with aggregating unstructured video data from thousands of cameras, and looking for specific events in real time? If this is the case then the data reduction will be crucial as tens if not hundreds of terabytes will be generated daily, and the fog nodes will need to make a critical decision (i.e. yes or no) based on this massive amount of data.
- **Number of Devices:** If the IoT or wearable device is simplistic (i.e. based on one sensor), then the generated data will be very small and a fog node is not needed. However, if the number of sensors changes depending on different situations, then the fog topology may need to scale up or down accordingly.
- **Reliability:** Forms of failure must be considered in IoT and wearable application. If one fog node fails, another has to fill in to perform the necessary action or service. This case is important in life-dependent applications that require real-time decision making (e.g. seniors remote health monitoring).

The most basic fog solution could be an edge processing unit (i.e. gateway, router) installed in close proximity to an array of sensors where a fog node can be used as a gateway to a WPAN network which communicates to a host.

Another topology may include the cloud as the parent over the entire fog network. Here, the fog node would aggregate data, enforce security, and perform the processing required to communicate with the cloud.

Multiple fog nodes can also be deployed which would be responsible for services and edge processing where each node is connected to a set of sensors. Each node is serviced by the cloud and has a unique identity in order to provide a unique set of services based on location.

Another model, which builds on the second one discussed in this section, can be established by enabling multiple fog nodes to communicate securely and privately to multiple cloud vendors. For example, in a smart city setting, a number of areas may be covered by different counties. Each county may favor one cloud vendor over the other, but all counties are required to use one approved security camera vendor. In this case, the camera vendor would have their single cloud instance coexist with multiple counties. Utilizing the referenced model, the fog nodes are able to steer data to multiple cloud service providers.

7.4 Platforms

A cloud platform for IoT and wearable technology is an essential component of their massive ecosystem. Since not all wearable devices need cloud services, we typically refer to such platforms as IoT platforms as an umbrella term.

An IoT platform is a multi-layer technology that facilitates provisioning, automation, and management of connected devices. It essentially connects a diversity of hardware to the cloud utilizing enterprise-grade security mechanisms, data processing capabilities, and connectivity options. An IoT platform provides a set of ready-to-use features for developers that could considerably speed up the development of applications for IoT and wearable devices, and cut down significant costs. Moreover, platforms are perfect when it comes to scalability and device heterogeneity.

Initially, IoT platforms were intended to act as a middleware, i.e. to function as a mediator between the hardware and application layers. To be practical, IoT middleware is expected to support interfacing with any type of connected devices and merge in with third-party applications without any issue.

Below are some of the popular IoT platforms available in the market today:

- **Amazon Web Services (AWS)**
  The cloud services provided by Amazon comprise an IoT suite that supports all aspects and needs of IoT applications. Examples of IoT services provided by AWS include AWS IoT Core, which deals with building IoT applications; AWS IoT Device Management which allows straightforward addition and organization of devices; AWS IoT Analytics, which provides a service for automated analytics of large amounts of diverse types of data that comes from different types of devices; and AWS IoT Device Defender, which supports the configuration of security mechanisms for connected devices.
- **Google Cloud IoT**
  Google Cloud is another global platform that supports IoT solutions. Its IoT package enables the developers to create and manage systems regardless of size and complexity. Dedicated IoT services include: Cloud IoT Core, Cloud Pub/Sub, and Cloud Machine Learning Engine.
- **Microsoft Azure IoT Suite**
  Microsoft Azure is another global cloud service provider in the same league as AWS and Google Cloud Platform. Azure IoT Suite offers both preconfigured and customizable solutions. Service packages similar to the ones offered by AWS and Google are available too.

Other major platforms include SAP, Salesforce IoT, Oracle Internet of Things, Cisco IoT Cloud Connect, IBM Watson Internet of Things, GE predix, Autodesk Fusion Connect, ThingWorx (now acquired by PTC), and Xively Platform.

Fog platforms also exist. For example, AWS IoT Greengrass extends AWS to edge devices so they can operate locally on the data they generate, while the cloud is still used for management, analytics, and archiving/storage purposes. With AWS IoT Greengrass, IoT and wearable devices can run AWS Lambda functions, use machine learning models, sync data in devices, and establish secure communication with other devices, even when they are not connected to the Internet.

With AWS IoT Greengrass, you can use familiar programming languages and models to create and test your device software in the cloud and then deploy it. AWS IoT Greengrass can be programmed using familiar languages and programming models to filter device data, manage the device data, and only send necessary information back to AWS. AWS IoT Greengrass Connectors can also be used to connect to third-party applications, on-premises software, and other AWS services.

AWS IoT Greengrass lets the developers create IoT solutions that connect different types of devices with the cloud and each other. Devices that run Linux, such as Raspbian, Arm, and x86 architectures can host AWS IoT Greengrass Core which enables the local execution of AWS Lambda code, security, messaging, and data management.

7.4.1 Criteria for Choosing a Platform

As seen from the previous section, there are numerous IoT platform options to choose from, which makes it difficult to find the best solution for the project in hand. Below are the major criteria for choosing an IoT platform:

- **Cost and Payment Model:** Some platforms use the pay-as-you-go model where the client is charged only for the resources they actually consume (i.e., AWS IoT Core), while other platforms use the subscription model with a flat fee monthly bill (i.e., Salesforce). Depending on the project needs, one should choose the payment model that works best.
- **Platform Stability:** With so many platforms in the market, it is likely that some will go down at some point. It is important to choose a platform from a reputable vendor that will likely be around for several years.
- **Platform Scalability and Flexibility:** In many cases, the project needs will change with time. Developers have to make sure that the chosen platform can accommodate the needs of the project if scaled up. In addition to scalability, the platform should be flexible enough to keep up with the newly emerging technologies, protocols, and features. Flexible platforms are often those that are built on open standards and are committed to keeping pace with the rapidly changing protocols and standards. It is also crucial that the platform is unbound to hardware and network.
- **Time to Market:** As mentioned previously, one of the greatest advantages of using an IoT platform is that it accelerates the time to market. A realistic estimate of how long the deployment process takes to get to market should be inquired prior to making the deal. Data analytics capabilities and data ownership are also important factors to consider when choosing a platform.

7.5 Data Analytics and Machine Learning

One of the core subjects in IoT and wearable technology is how to make sense of the massive amount of data that is generated. As mentioned previously, the real impact of data coming from smart devices is realized only when the analysis of the data leads to actionable business insights.

Because much of this data can seem beyond grasp, specialized algorithms and tools are needed to find the data relationships that will lead to useful insights. This brings us to the topic of machine learning.

Machine learning is part of a larger set of technologies commonly grouped under the umbrella of artificial intelligence (AI). Once collected data are analyzed, intelligent actions need to be taken. Performing such analysis manually is close to impossible or very impractical.

The most useful feature of machine learning in IoT and wearable technology is that it can detect outliers and abnormal activities and trigger necessary actions accordingly. As it learns more and more about an event or activity, it gets more accurate and efficient. Moreover, machine learning algorithms can create models which predict future events precisely by identifying the factors that lead to a particular result.

The difficulty, however, lies in determining the right algorithm and the most appropriate learning model for each use case. Such analysis goes beyond the scope of this chapter, and the reader is referred to a couple of resources that can be found in the references section of this chapter ([2, 3]).

7.6 Conclusion

The cloud offers a wide range of functions and solutions; however, certain analysis should be performed on the edge, closer to the devices where data are being generated to solve security, cost, and latency issues found in cloud.

Selecting the cloud service models and frameworks, fog topology, and analytics modules is an important task where much literature dives deep into the minute details of creating and deploying them. The design team must have an understanding of the topology and framework and be able to choose the best architecture to address the data needs of the project in hand with possible future scaling in mind.

It was demonstrated in this chapter that the maximum gain from IoT or wearable technology projects can only be achieved from an optimal combination of cloud and edge computing where both work together to achieve the desired outcomes of the project.

Problems

1. Research examples of IoT and wearable devices with each example utilizing one of the cloud types mentioned in this chapter.
2. Research examples of IoT and wearable devices with each example utilizing one of the cloud service models mentioned in this chapter.
3. You are working on developing a new smart home virtual assistant. Research five of the IoT platforms mentioned in this chapter, then narrow down your selection for this project to two candidates. Justify your choices.
4. List five applications or device ideas that could benefit from a fog layer. Justify your answer.
5. Research five open-source API platforms and compare between their functionalities, features, and related criteria in a comparison table.
6. Comment on the mechanism of a simple REST client example for retrieving API data from an IoT or wearable device.
7. Pick an IoT or wearable technology application and comment on how involving a machine learning algorithm will lead to more useful insights.
8. Compare through a table the differences between the OpenStack and OpenFog architectures

Technical Interview Questions

1. What are the main service models in cloud computing?
2. What would be your potential choices of platforms to be used in a project based on large-scale cloud computing?
3. What are some of the cloud computing platform databases used in the industry? What would your choice be for a fitness tracking project?
4. Describe the process of cloud migration.
5. How would you move workloads to the cloud using Azure?

6 Why Application Programming Interfaces (APIs) are used in cloud services?

7 What are some of the important questions you would ask a client if you were tasked with migrating a high traffic on-premise data and application to the cloud?

8 How would you move 1 billion files from an on-premise data center to the cloud?

9 How have you used DevOps[3] in a project and how has it added value to your clients?

10 Describe a deployment of cloud computing for an IoT project.

11 Describe cloud application security requirements.

12 How are XML and JSON used in cloud computing?

13 Comment on the importance of a database in Edge computing.

14 What are the three types of data used in cloud computing?

15 What are the cloud architecture layers used in AWS?

16 What is the difference between elasticity and scalability in cloud computing?

References

1 OpenFog Consortium (2017). OpenFog reference architecture for fog computing, Produced by the OpenFog Consortium Architecture Working Group. https://www.openfogconsortium.org/wp-content/uploads/OpenFog_Reference_Architecture_2_09_17-FINAL.pdf.
2 Kapoor, A. (2019). *Hands-on Artificial Intelligence for IoT: Expert Machine Learning and Deep Learning Techniques for Developing Smarter IoT Systems*. Birmingham, UK: Packt.
3 Al-Turjman, F. (2019). *Artificial Intelligence in IoT*. Lüneburg, Germany: Springer.

3 DevOps is a set of practices that integrate software development and IT operations aiming at shortening the project development life cycle.

Further Reading

Accenture (2010). Cloud computing and sustainability: the environmental benefits of moving to the cloud, Accenture, Dublin.

Binz, T., Breiter, G., Leyman, F., and Spatzier, T. (May 2012). Portable Cloud Services Using TOSCA. *IEEE Internet Computing* 16 (3): 80–85.

Binz, T., Breitenbücher, U., Kopp, O. et al. (2013). Improve resource-sharing through functionality- preserving merge of cloud application topologies. *Proceedings of the 3rd International Conference on Cloud Computing and Service Science, CLOSER 2013*. Aachen, Germany: SciTePress.

Breitenb Ucher, U., Binz, T., Kopp, O., and Leymann, F. (2013). Pattern-based Runtime management of composite cloud applications. *Proceedings of the 3rd International Conference on Cloud Computing and Service Science, CLOSER 2013*. Aachen, Germany: SciTePress.

Chen, X., Jiao, L., Li, W., and Fu, X. (2016). Efficient multi-user computation offloading for mobile-edge cloud computing. *IEEE/ACM Transactions on Networking* 24: 2795–2808.

Coraid (2013). The fundamentals of software-defined storage – Simplicity at scale for cloud architectures. A white paper by Coraid.

Grindle, M., Kavathekar, J., and Wan, D. (2013). A new era for the healthcare industry - Cloud computing changes the game. A white paper by Accenture.

Khan, A.M. and Freitag, F. (2017). On edge cloud service provision with distributed home servers. *Proceedings of the IEEE International Conference on Cloud Computing Technology and Science (CloudCom)*, Hong Kong (11–14 December 2017).

Li, F., Ogler, M.V., Claeß ens, M., and Dustdar, S. (2013). Efficient and scalable IoT service delivery on cloud. *6th IEEE International Conference on Cloud Computing, (Cloud 2013)*, Industrial Track, Santa Clara, CA, USA.

Li, F., Vögler, M., Sehic, S. et al. (2013). Web-scale service delivery for smart cities. *IEEE Internet Computing* 17 (4): 78–83.

Lin, J., Yu, W., Zhang, N. et al. (2017). A survey on internet of things: architecture, enabling technologies, security and privacy, and applications. *IEEE Internet of Things Journal* 4: 1125–1142.

Liu, H., Eldarrat, F., Alqahtani, H. et al. (2017). Mobile edge cloud system: architectures, challenges, and approaches. *IEEE Systems Journal* 12 (3): 1–14.

Miluzzo, E. (2014). AT&T labs research, I'm cloud 2.0, and I'm not just a data center. *IEEE Computer Society* 18 (03): 73–77.

OASIS (2013). Topology and orchestration specification for cloud applications (TOSCA).

Östberg, P.O., Byrne, J., Casari, P. et al. (2017). Reliable capacity provisioning for distributed cloud/edge/fog computing applications. *Proceedings of the 2017*

European Conference on Networks and Communications (EuCNC), Oulu, Finland (12–15 June 2017), pp. 1–6.

Ren, J., Pan, Y., Goscinski, A., and Beyah, R.A. (2018). Edge computing for the internet of things. *IEEE Network* 32: 6–7.

Satyanarayanan, M. (2017). The emergence of edge computing. *Computer* 50: 30–39.

Simmhan, Y. (2013). Cloud-based software platform for big data analytics in smart grids. *Co-published by the IEEE CS and the AIP*, California, USA: University of Southern California.

Wang, W. (2012). Integrating sensors with the cloud using dynamic proxies. *IEEE 23rd International Symposium on Personal Indoor and Mobile Radio Communications (PIMRC)*, Sydney, Australia.

Wang, L., Jiao, L., Kliazovich, D., and Bouvry, P. (2016). Reconciling task assignment and scheduling in mobile edge clouds. *Proceedings of the IEEE 24th International Conference on Network Protocols (ICNP)*, Singapore (8–11 November 2016), pp. 1–6.

Wang, N., Varghese, B., Matthaiou, M., and Nikolopoulos, D.S. (2017). ENORM: a framework for edge node resource management. *IEEE Transactions on Service Computing* Sustainability 2018: 10.

Wettinger, J., Behrendt, M., Binz, T. et al. (2013). Integrating configuration management with model-driven cloud management based on TOSCA. *Proceedings of the 3rd International Conference on Cloud Computing and Service Science, CLOSER 2013*. Aachen, Germany: SciTePress, pp. 437–446.

Xu, L. (2012). Cloud-based monitoring framework for smart home. *IEEE 4th International Conference on Cloud Computing Technology and Science*, Taipei, Taiwan.

Xu, J., Palanisamy, B., Ludwig, H., and Wang, Q. (2017). Zenith: utility-aware resource allocation for edge computing. *Proceedings of the IEEE International Conference on Edge Computing (EDGE)*, Honolulu, HI, USA (25–30 June 2017), pp. 47–54.

Yerva, S.R. (2012). Cloud-based social and sensor data fusion. *The International Conference on Information Fusion*, Singapore.

Zhang, Y., Huang, H., Xiang, Y. et al. (2017). Harnessing the hybrid cloud for secure big image data service. *IEEE Internet of Things Journal* 5 (5): 1380–1388.

Zhu, Q., Wang, R., Chen, Q. et al. (2010). IOT gateway: bridging wireless sensor networks into internet of things. *2010 IEEE/IFIP International Conference on Embedded and Ubiquitous Computing*, Hong Kong, China, pp. 347–352. [4] Tridium, "JACE Controller.".

8

Security

8.1 Introduction

While many of the emerging IoT and wearable technology applications are giving rise to beneficial and innovative uses, they also pose security and privacy concerns that are largely unexplored. In fact, a new research area concerning the security of these technologies has recently emerged.

According to a recent study on ten commercially available smartwatches, it was found that all of them exhibit a form of vulnerability which includes poor authentication, weak encryption techniques, and privacy-related issues. For instance, only 50% of the surveyed devices offered a screen lock mechanism by either a PIN or a custom pattern, while 70% transmitted data without any form of encryption. It was recently reported that a hacker could control the mechanism of an insulin regulator wirelessly from hundreds of feet away to deliver a lethal dose to a user. Further, it was demonstrated that a hacker could deliver a lethal voltage shock to a patient with a pacemaker. One can think of other horrifying scenarios!

The infrastructure of IoT and wearable systems encompass a wide spectrum of components and technologies, and each is susceptible to a number of vulnerabilities and threats. It is crucial to ensure that each component is safe and secure.

Compared to laptops, smartphones, and tablets, which were swiftly embraced by consumers, IoT and wearable devices are being adopted on a relatively slower pace. However, it is never too early to pay particular attention to the inevitable security risks that come with these technologies. These concerns will be more serious in the coming years as these devices become more mainstream, and without the right security controls, data exchanged and shared by IoT and wearable devices could end up being used in ways never intended or even imagined. New forms of identity theft, harassment, stalking, and fraud are already emerging.

Fundamentals of IoT and Wearable Technology Design, First Edition. Haider Raad.

Published 2021 by John Wiley & Sons, Inc.

This Chapter examines the security goals that every designer should aim to achieve. Next, an overview of the most important security challenges, threats, attacks, and vulnerabilities faced by IoT and wearable devices is provided. Finally, a list of security design consideration and best practices that have historically worked are discussed.

8.2 Security Goals

The goals of information security are defined best by the CIA triad. Not to be confused with the U.S government agency, CIA stands for Confidentiality, Integrity, and Availability. The goals of information security, and its largest branch, cybersecurity, are to protect the confidentiality of information, the integrity of information from unauthorized changes, and to ensure the availability of information to the users at the expected performance level. The term information and data in the CIA triad has a broad definition, and it spans from high-level user information to metadata clues.

With IoT and wearable solutions maturing and taking on key responsibilities, security becomes a critical issue. Like everything else, makers of IoT and wearable devices are forced to develop their products within the boundaries of the information security goals which apply to both data in motion (during transmission) or data at rest (stored data or system configuration). However, these devices create new entry points for attackers to get into the system which requires additional security goals on top of the CIA:

Confidentiality: Confidentiality is an important security feature in IoT, but it may not be mandatory in some scenarios where data is presented publicly. However, in most cases and scenarios data is sensitive and must not be disclosed or read by unauthorized entities. Sensitive data include, but not limited to, personal biometric data from wearables and smart home applications, patient data, private business data, and military data.

Integrity: Integrity models ensure that data remain clean and trustworthy by protecting against intentional or accidental changes to the system data.

Availability: Availability models keep data, services, and resources available for authorized users at any given time or situation (i.e. during natural disasters). Different hardware and software components in IoT and wearable devices must be able to provide services even in malicious environments or adverse situations. Different applications have different availability requirements, for example, remote health monitoring systems would most likely have higher availability requirements than a farm soil moisture monitoring system.

Authentication and Authorization: Authentication is the process of identifying the device and confirming the device's ID is valid, while authorization provides a mechanism to associate a specific device to certain permissions. Different authentication and authorization requirements call for different solutions in different applications.

Auditing: Auditing refers to the systematic evaluation of the security of a device by assessing how well it performs when it is tested under an established criteria. In IoT and wearable devices, the need for auditing depends on the type of application and its value. IoT auditing services are actually offered by a number of businesses in the industry where every feature in the device is tested for a list of security vulnerabilities.

Nonrepudiation: Nonrepudiation serves as a definitive proof of the validity and origin of all data transmitted and received. In other words, it ensures that a sender cannot deny having sent the data, nor the receiver can deny having received the data. It is worth noting that this has limited applications in IoT and wearable technology.

8.3 Threats and Attacks

The sophistication of cyberattacks targeting IoT and wearable devices is on the rise. In fact, 2016 witnessed the emergence of an IoT-based botnet[1] that almost paralyzed the Internet.

Besides botnets, the following years witnessed a growing increase in malwares, and cryptominers that can be used to target cryptocurrency. It is also worth mentioning that cloud storage poses threats with high impact due to the large amount of Personally Identifiable Information (PII). Moreover, attacks on the cloud are normally sophisticated, coordinated, planned well ahead of time and are conducted by professional cyber attackers.

The motivations behind cybercrimes can be quite simple: money and information. According to one study, financial and espionage-driven motivations make up about 93% of attacks. Attacks driven by large-scale disruption and destruction such as terrorism and revenge attacks are among other motivations.

Some of the most notable attacks including malware based (such as Mirai, Satori, and VPNFilter), exploit kits, and advanced persistent worms such as Stuxnet. Cybercriminal and state-sponsored activities which are exploited to access sensitive data, instill operational, or cause a reputational damage are possible due to system vulnerabilities such as:

---

1 **A botnet** is a number of devices connected to the Internet that can be used to perform Distributed Denial-of-Service (DDoS) attacks, access device data, send spam, etc.

a) Weak authentication mechanisms
b) Weak password standards
c) Poor cryptography implementation which promotes man-in-the middle attacks, and session and protocol hijacking

8.3.1 Threat Modeling

To better illustrate the threat landscape in the realm of IoT and wearable technology, we can use threat modeling.

Threat modeling is a structural and systematic way by which potential threats can be identified, prioritized, and eventually mitigated. The objective of such modeling is to provide cybersecurity teams with a systematic perspective of the potential attacker's profile, attack vectors, and the assets most likely to be targeted by an attacker.

There are various ways and methodologies of establishing threat models, we will adopt one model created by Microsoft, called *STRIDE*:

Spoofing: A spoofing attack takes place when an attacker pretends to be someone they're not. An attacker may be able to pull out cryptographic key data from a device, then accesses the system with another device using the stolen identity of the device the key has been taken from.

Tampering: Tampering refers to maliciously modifying processes, data at rest, or data in-transit. An attacker may partially or fully replace the software running on the device, which can potentially have adverse consequences if the attacker is able to add or remove some functional elements, or modify or destroy important data.

Repudiation: Repudiation refers to denying that an activity or an event has taken place. Attackers often try to hide their malicious actions, to avoid getting detected. For example, they might try to erase their illegal activities from the logs or spoof the credentials of another user.

Information Disclosure: Information Disclosure refers to data leaks or data breaches. Attackers may try to run a modified software on the compromised system which could potentially help leak data to unauthorized parties. In IoT and wearable applications, the attacker may try to gain access into the communication path between the device and gateway, or gateway to cloud to extract information.

Denial of Service: Denial of Service refers to degrading or denying a service or network resources to users. In some cases, attackers would benefit from preventing users to access a system, for example as a way to blackmail and coerce them.

Elevation of Privilege: Elevation of Privileges refers to gaining access to resources that one is not allowed. Once a user is identified on a system, they typically have some form of privileges, for instance, they are authorized to perform some actions or access some resources, but not necessarily all of them. Therefore, an attacker might attempt to gain additional privileges, for example by spoofing a user with higher privileges or by tampering the system to upgrade their own privileges.

8.3.2 Common Attacks

Common cyberattack types in IoT and wearable technology based on such threat models include:

1) **SQL Injection Attacks:** An attacker can use an SQL (Structured Query Language) injection to bypass the authentication and authorization mechanisms of a web application mainly to access the contents of a database. It can also be used to modify the records in a given database (which affects data integrity) and to provide an attacker with unauthorized access to sensitive information including: PII, which is information that can be used to identify, contact, or locate a user, customer data, and other sensitive data.
2) **Account Brute Force Login Attacks**[2]**:** This is one of the oldest and most common attacks performed against Web applications. The goal of a brute force attack is to penetrate user accounts by repeatedly attempting to guess the password of a user.
3) **Distributed Denial-of-Service (DDoS) Attacks:** a DDoS attack is an attempt to force an online service to be unavailable via overwhelming it with excessive traffic. Due to low computational power and memory capabilities, the majority of IoT devices are vulnerable to such attacks.
4) **Default Password and Dictionary Attacks:** are the most common large-scale attacks. Failing to change the vendor's default password is a substantial security risk. It is worth noting that some computing devices, such as routers, typically come with a unique default password printed on a sticker (affixed on the back of the device), which is considered a more secure option. However, some manufacturers derive the password from the device's MAC address using a standard algorithm, in which case passwords could be easily reproduced by cybercriminals. Dictionary attacks, on the other hand, are performed by trying possible combinations of letters and numbers to guess the password.
5) **Back Door Attacks:** A backdoor attack is a way of accessing a system program through bypassing its security mechanisms. A code developer may intentionally install a back door so that the program can be accessed for troubleshooting

2 **Brute Force Attack:** is a trial and error technique used by application algorithms to decode encrypted data such as passwords or Data Encryption Standard keys.

purposes. However, hackers often make use of back doors as a part of a malicious exploit.

6) **Physical Attacks and Theft:** This type of attack deals with tampering with the hardware components. IoT devices are more susceptible to such attacks than wearables due to their unattended and geographically distributed nature. Moreover, IoT devices operate in outdoor environments in many applications which make them more prone to tampering and theft.

8.4 Security Consideration

Security for IoT devices has to be considered from the first phase of design and not retrofitting it at the end since it would be too late at that point. Security also needs to be viewed both holistically and atomistically from the "thing" to the cloud. Each component in the system should have a list of security parameters and enablers.

A plethora of lists of security considerations and ideas that have traditionally worked exist in the literature. For example, enabling safe over-the-air updates to maintain security codes up to date, securing sensitive data and safeguarding regulatory compliance, managing the lifecycle of each device, and using the most current operating system and libraries with all relevant patches have all been recommended as good practices during the design phase.

Biometric user authentication has also been considered as a possible solution to some common threats. Biometric authentication offers more convenience in wearable devices, especially in compact platforms where passwords and PIN codes would be less appropriate. However, this solution will cover specific aspects of the raised privacy and security concerns, but could certainly trigger more.

In addition to the functional aspects, security solutions for IoT and wearable technology have to also be scalable and flexible enough to be integrated with platforms of enterprise systems in cost-effective manner.

In 2019, the European Union Agency for Cybersecurity (ENISA) published a document entitled "Good Practices for Security of IoT." The aim of the report was to provide guidelines and recommendations for designers and developers for countering and mitigating the threats impacting IoT. The report recognizes that securing IoT can be a tough task for software developers if hardware is not equipped with security capabilities. For instance, when integrating a powerful cryptographic algorithm in the software stack, a Trusted Platform Module (TPM) is used in the hardware to ensure that the private key will not be compromised. Therefore, the underlying hardware cannot be neglected when developing IoT projects. In fact, the security approach is conceived as a set of requirements where the design of hardware influences the design of software.

A comprehensive view of the Secure IoT Software Development Life Cycle (SDLC) landscape which indicates the areas that require protection are classified into three main groups:

- **People:** Security considerations that could impact every stakeholder involved in the life cycle of a product, from the developer(s), to the end users.
- **Processes:** This involves all security aspects of the software conception, development, to deployment in the market.
- **Technologies:** This involves technical procedures and elements used to reduce vulnerabilities and deficiencies during the software development process.

In this section, we will list the measures that are most relevant to the design of IoT and wearable devices.

1) Test the Third Party Process: Prior to integrating components or services from third party suppliers, a process must be defined to test their security performance.
2) Define an Incident Management Plan: A plan to manage vulnerabilities and updates must be defined, including third-party components, along with the necessary actions to combat security incidents during software development.
3) Implement Configuration Management and Authorization Policy: The integrity of the system has to be appropriately managed by ensuring that only authorized changes can be made to the configuration. A privilege based scheme has to also be established to prevent unauthorized users from accessing restricted resources.
4) Define Security Metrics: Security metrics have to be defined and implemented to ensure that the security requirements are fulfilled throughout the lifecycle.
5) Provide a Secure Framework: A framework to implement security by design must be defined and implemented throughout the solution lifecycle.
6) Specify Security Requirements: To include features that ensure regulatory compliance and avoid vulnerabilities; security requirements must be identified prior to development.
7) Perform Risk Assessment: Risks throughout the software development process have to be identified by analyzing all the data sources, storage, applications, or third parties, if any.
8) Implement Data Classification: Data has to be classified based on their level of sensitivity to establish appropriate protection measures.
9) Ensure That the Hardware Requirements Derived from Software Requirements are Considered: Additional requirements stemming from hardware implementation must be defined and documented.
10) Implement Authorization: In an IoT system, access control must be implemented to verify that users and applications have the right permissions.

11) Secure Storage of Users' Credentials: User credentials of IoT devices must be protected from disclosure, i.e. using hash functions for storing passwords.
12) Use Libraries and Third-Party Components that are Patched for Latest Known Vulnerabilities: Software libraries and frameworks to be included in the project must be verified that they are patched for the latest known security vulnerabilities. An upgrade roadmap for libraries and third-party components must also be established.
13) Use Secure Communication Protocols: It must be ensured that communications will not be compromised by utilizing encrypted channels and authenticated connections in order to share data between IoT devices.
14) Use Proven Encryption Techniques: System data should be protected using encryption algorithms that are proven to be secure.
15) Implement Secure Web Interfaces: Web interfaces or technologies used in IoT systems should also be secured in order to be used.
16) Implement Secure Coding Practices: During the design process, the software under development must be tested to ensure that the authentication mechanism conforms to globally accepted practices and that queries use parameterization to avoid code injections.
17) Implement Anti-Tampering Features: Countermeasures have to be deployed to prevent unauthorized code modification in all steps of the development process.
18) Perform IoT SDLC Tests: A penetration test has to at least be carried out when the software development is complete.
19) Enforce the Change of Default Settings: Change of default settings must be ensured at first user interaction with the device.
20) Use Substantiated Underlying Components: Component customizations must be restricted in order to not compromise security functionalities.
21) Implement Interoperability Open Standards: This is done to enhance secure integration processes.
22) Provide Audit Capability: Security events must be ensured that they are registered in software logs.

8.4.1 Blockchain

A blockchain is a series of time-stamped records of data that are linked and secured using cryptography. Each block contains a cryptographic hash of the previous block, and transactions are managed by a cluster of computers not owned by any single entity.

Characteristics of blockchain are:

1) Data contained in a blockchain is resistant to modification.
2) Transactions between two parties are recorded permanently and verifiably by an open, distributed ledger.

3) A blockchain is typically managed by a peer-to-peer network conforming to a protocol for inter-node communication that constantly validates new blocks.
4) Once the data contained in a block is recorded, it cannot be altered retroactively without altering the entirety of all subsequent blocks, which requires consensus from the majority in a given network.

Blockchain nodes are somewhat similar to how smart objects and systems are connected in a network. It treats the data transaction the same way it would treat financial transactions on a Bitcoin network, and hence, IoT and wearable systems utilizing blockchain would allow secure, consensus-based messaging between nodes in a network. This will lead to simplifying business processes, improving transparency, providing autonomy, saving costs, and making connected objects such as cars and appliances more reliable and secure.

Here is how blockchain and smart devices can work together:

- **Security:** The ledger used in blockchain cannot be manipulated or tampered with which adds another layer of security if implemented in an IoT or wearable system. Moreover, the autonomous security solution blockchain provides makes it a perfect element for IoT and wearable solutions.
- **Decentralization:** The power of a blockchain lies in the fact that there is no single entity controlling the state of transactions. Furthermore, redundancy is enforced in the system by ensuring that every node using blockchain maintains a copy of the ledger. Assuming trustless messaging between nodes in a blockchain network, the system must live by consensus.
- **Encryption and Distribution:** The use of encryption and storage distribution in blockchain allow data to be recorded securely in IoT and wearable systems without any human interference which preserves the data integrity allowing it to be trusted by all parties involved in the network.
- **Communication Assurance:** Blockchain allows IoT and wearable devices to securely communicate and exchange transactions with a very high assurance that everything will be processed as per the predefined terms of contract.
- **Cost Saving:** Since no middleman is needed in exchanged and shared blockchain data, significant costs can be saved in the transaction chain. Using smart contracts, blockchain can allow IoT and wearable devices to automate data transactions across various networks.
- **Tracking:** Blockchain can keep unalterable records of the history of an IoT or wearable device. In a network, this property would allow smart devices to autonomously function without the need for a centralized authority. This is just like in cryptocurrencies where direct payment services are provided without the need of any third-party handler.

Blockchain and IoT are two technologies that will continue to thrive as they both become more pervasive and mainstream.

The aim of this section was to introduce blockchain technology, and there is certainly much more material to cover but it is beyond the scope of this book. The reader is referred to some suggested books specialized in this vital topic (in the Further Reading section). It is crucial that one understands the complexities involved in blockchain deployment before launching into using it for IoT or wearable projects.

8.5 Conclusion

IoT and wearable technology face a number of threats that must be recognized for proactive measures to be taken. These technologies comprise a conglomeration of components, and each of these components is subjected to a number of vulnerabilities and threats. It is crucial to ensure that each component is safe and secure.

Security is a process that must be considered in the first phase of design and applied throughout the lifecycle of product. To achieve security, the design and development team must be able to define risks and make informed decisions about how to best address them. Fortunately, much of what is needed to minimize risks from threats is already available. Networks can be secured with the right equipment, configuration, practices, and policies. Threats and attacks from risky practices of unaware users can be identified and mitigated with the right techniques and training.

Problems

1 How would cybersecurity affect the development and implementation of the IoT and wearable technology globally?

2 Research the commercially available security solutions dedicated for IoT and wearable technology. Compare between their effectiveness based on the targeted area.

3 You are working on prototyping a smart garden moisture sensor. Assuming you are still in the early stages of the prototype (breadboarding), what would you do to secure a) software, b) hardware?

4 Create a table that maps the security goals against threats and attacks mentioned in this chapter.

5 You are designing a wearable device that controls a pacemaker. What are the possible threats and attacks? What is your risk assessment strategy?

6. You are working on a project that involves designing a smart door lock that can be controlled remotely. What are most important security considerations?

7. How would you prevent brute force attacks when planning your IoT projects?

8. Research how Mirai and Satori attacks are related. What can you do to prevent such attacks in the future?

9. Research the different deployment models a DDoS mitigation provider may offer. What would you choose for an IoT-based smart home project?

10. Would your soil moisture monitor project benefit from blockchain technology? Why, or why not?

Technical Interview Questions

1. What is the difference between encoding and encryption?

2. What is the difference between encryption and hashing?

3. What are some of the most common cyberattacks?

4. Talk about cognitive cybersecurity?

5. What is SQL injection? How would you prevent it?

6. How would you detect IoT security incidents on your product?

7. What are the differences between Vulnerability Assessment and Penetration Testing?

8. What is the difference between Intrusion Detection System and Intrusion Prevention System?

9. What is the difference between symmetric and asymmetric encryption? Which one would you use for an IoT or wearable project?

10. What are the main assets you would focus on in your security strategy for developing an IoT project?

Further Reading

Abomhara, M. and Køien, G.M. (2015). Cyber security and the internet of things: vulnerabilities, threats, intruders and attacks. *Journal of Cyber Security and Mobility* 4 (1): 65–88.

Andreev, S. and Koucheryavy, Y. (2012). Internet of things, smart spaces, and next generation networking. *Proceedings of 12th International Conference, NEW2AN 2012, and 5th Conference, ruSMART 2012*, St. Petersburg, Russia (27–29 August 2012). LNCS, Springer, vol. 7469, 464.

Ansari, S., Rajeev, S., and Chandrashekar, H. (2002). Packet sniffing: a brief introduction. *IEEE Potentials* 21 (5): 17–19.

Ball, D. (2011). Chinas cyber warfare capabilities. *Security Challenges* 7 (2): 81–103.

Bergman, N. and Rouse, J. (2013). *Hacking Exposed Mobile: Security Secrets & Solutions*, 1e. New York: McGraw-Hill.

Bertino, E., Martino, L.D., Paci, F., and Squicciarini, A.C. (2010). Web services threats, vulnerabilities, and countermeasures. In: *Security for Web Services and Service-Oriented Architectures* (eds. E. Bertino, L.D. Martino, F. Paci and A.C. Squicciarini), 25–44. Berlin, Germany: Springer.

Brauch, H.G. (2011). Concepts of security threats, challenges, vulnerabilities and risks. In: *Coping with Global Environmental Change, Disasters and Security* (ed. H.G. Brauch), 61–106. Berlin, Germany: Springer.

Cha, I., Shah, Y., Schmidt, A.U. et al. (2009). Trust in M2M communication. *IEEE Vehicular Technology Magazine* 4 (3): 69–75.

Dahbur, K., Mohammad, B., and Tarakji, A.B. (2011). A survey of risks, threats and vulnerabilities in cloud computing. *Proceedings of the 2011 International Conference on Intelligent Semantic Web-Services and Applications*. Amman, Jordan: ACM, 12.

De, S., Barnaghi, P., Bauer, M., and Meissner, S. (2011). Service modelling for the internet of things. *2011 Federated Conference on Computer Science and Information Systems (FedCSIS)*. Szczecin, Poland: IEEE, pp. 949–955.

Dolev, D. and Yao, A.C. (1983). On the security of public key protocols. *IEEE Transactions on Information Theory* 29 (2): 198–208.

Duncan, A.J., Creese, S., and Goldsmith, M. (2012). Insider attacks in cloud computing. *2012 IEEE 11th International Conference on Trust, Security and Privacy in Computing and Communications (TrustCom)*. Liverpool, England, UK: IEEE, pp. 857–862.

ENISA (2019). Good Practices for Security of IoT, Secure Software Development Lifecycle, The European Union Agency for Cybersecurity (ENISA).

FTC Staff Report (2015). Internet of Things, Privacy & Security in a Connected World. FTC Staff Report.

Hongsong, C., Zhongchuan, F., and Dongyan, Z. (2011). Security and trust research in M2M system. *2011 IEEE International Conference on Vehicular Electronics and Safety (ICVES)*. Beijing, China: IEEE, pp. 286–290.

Rohan, P. (2019). Demystifying the relationship between IoT and blockchain. *Forbes* (29 May 2019). https://www.forbes.com/sites/forbestechcouncil/2019/05/29/demystifying-the-relationship-between-iot-and-blockchain/#267f0c8d605d.

Jiang, D. and ShiWei, C. (2010). A study of information security for M2M of IoT. *2010 3rd International Conference on Advanced Computer Theory and Engineering (ICACTE)*, vol. 3. Chengdu, Sichuan province, China: IEEE, pp. V3–V576.

Ke, W.C. and Singh, M.M. (2016). Wearable technology devices security and privacy vulnerability analysis. *International Journal of Network Security & Its Applications (IJNSA)* 8 (3): 19–30.

Kizza, J.M. (2013). *Guide to Computer Network Security*. Heidelberg, Germany: Springer.

Kozik, R. and Choras, M. (2013). Current cyber security threats and challenges in critical infrastructures protection. *2013 Second International Conference on Informatics and Applications (ICIA)*. Lodz, Poland: IEEE, pp. 93–97.

Kumar, J.S. and Patel, D.R. (2014). A survey on internet of things: security and privacy issues. *International Journal of Computer Applications* 90 (11): 20–26, published by Foundation of Computer Science, New York, USA.

Li, F., Lai, A., and Ddl, D. (2011). Evidence of advanced persistent threat: a case study of malware for political espionage. *2011 6th International Conference on Malicious and Unwanted Software (MALWARE)*. Fajardo, USA: IEEE, pp. 102–109.

Lopez, J., Roman, R., and Alcaraz, C. (2009). Analysis of security threats, requirements, technologies and standards in wireless sensor networks. In: *Foundations of Security Analysis and Design V* (eds. J. Lopez, R. Roman and C. Alcaraz), 289–338. Berlin, Germany: Springer.

Padmavathi, D.G. and Shanmugapriya, M. (2009). A survey of attacks, security mechanisms and challenges in wireless sensor networks. arXiv preprint arXiv:0909.0576.

Pipkin, D.L. (2000). *Information Security*. Saddle River, NJ: Prentice Hall PTR.

Roman, R., Zhou, J., and Lopez, J. (2013). On the features and challenges of security and privacy in distributed internet of things. *Computer Networks* 57 (10): 2266–2279.

Rudner, M. (2013). Cyber-threats to critical national infrastructure: an intelligence challenge. *International Journal of Intelligence and Counter Intelligence* 26 (3): 453–481.

Santos, M. and Moura, E. (2019). *Hands-On IoT Solutions with Blockchain: Discover How Converging IoT and Blockchain Can Help You Build Effective Solutions*. Birmingham, UK: Packt Publishing Ltd.

Schneier, B. (2011). *Secrets and Lies: Digital Security in a Networked World*. New Jersey: John Wiley & Sons.

Stango, A., Prasad, N.R., and Kyriazanos, D.M. (2009). A threat analysis methodology for security evaluation and enhancement planning. *2009 Third International Conference on Emerging Security Information, Systems and Technologies. SECURWARE'09*. Athens/Glyfada, Greece: IEEE, pp. 262–267.

Taneja, M. (2013). An analytics framework to detect compromised IoT devices using mobility behavior. *2013 International Conference on ICT Convergence (ICTC)*. IEEE, pp. 38–43.

Vermesan, O., Friess, P., Guillemin, P. et al. (2011). Internet of things strategic research roadmap. *Internet of Things-Global Technological and Societal Trends* 7858: 9–52.

Watts, D. (2003). Security and vulnerability in electric power systems. *35th North American Power Symposium*, Wichita, Kansas, USA, vol. 2, pp. 559–566.

Zhu, L., Gai, K., and Li, M. (2019). *Blockchain Technology in Internet of Things*. Germany: Springer.

9

Concerns, Risks, and Regulations

9.1 Introduction

In 2015, a father walked into his three-year-old son's room, hearing a voice of an adult male coming through the baby monitor, saying "Wake up little boy, daddy's looking for you." The kid's family found out that the baby monitor had been remotely hacked by a stranger, which was also able to control the camera of the baby monitor and spy on the family.

While IoT and wearable technology are giving rise to a spectrum of new applications and innovative uses, as well as promising super attractive user benefits, they also pose new concerns that are largely unexplored.

This chapter first addresses the privacy issues and concerns arising from IoT and wearable technology, including those related to health data and data collected from children. The chapter next turns to safety and health issues, then discusses social and psychological impacts of these technologies. Finally, the chapter examines regulatory actions in the United States set by the federal government, including the Federal Trade Commission (FTC), National Telecommunications and Information Administration (NTIA), and by private companies practicing self-regulation within the industry. As a means of comparison, this chapter next discusses the regulatory actions taken by the European Union.

9.2 Privacy Concerns

The topic of privacy concerns in technology is not new, and it is certainly not limited to digital technologies. In fact, it has been discussed as early as in the nineteenth century by Louis Brandeis and Samuel Warren who defined in their work "The Right to Privacy" the protection of the private domain as the founding basis

Fundamentals of IoT and Wearable Technology Design, First Edition. Haider Raad.

Published 2021 by John Wiley & Sons, Inc.

of individual freedom in the modern age in response to the capacity elevation of government, press, and their related institutions to invade facets of personal activities that became accessible under new technological change.

However, privacy issues related to mobile and emerging technologies are relatively new, and complex to study and analyze. Furthermore, IoT and wearable devices that continuously collect information utilizing peripherals and sensors such as: microphones, cameras, and Global Positioning Systems, add newer challenges to the user's privacy.

Compared to laptops, smartphones, and tablets, which were swiftly embraced by consumers, IoT and wearable devices are being adopted on a relatively slower pace. However, it's never too early to pay particular attention to the inevitable privacy risks that they will bring. New forms of identity theft, harassment, stalking, and fraud are already emerging.

Previous studies show that most users are not aware of *what* data are being collected and *how* they are processed. For example, whenever an Apple phone user asks Siri a question, their voice is sent as a file to the data servers at Apple's headquarter for analysis. The file is then given a number that associates the phone to the question asked. These data are then stored on Apple servers for up to two years for testing purposes, though the file number is deleted after six months.

The FTC has addressed the privacy concerns associated with IoT devices due to the fact that they collect a large amount of sensitive information, including biometric data, geolocations, and financial information. The FTC has also reported that if 10000 households use an IoT-based home automation product from a single company they can collectively generate 150 million data points each day. In 2014, the FTC Bureau of Consumer Protection addressed the potential privacy risks of geolocation data collected from wearables and discussed how location data could potentially expose highly personal information about an individual, such as whether a user has visited an AIDS clinic, a hospital, or a worship facility. Clearly, such data can be misused if accessed by malicious attackers, if traded with companies, or if gathered by stalking apps. Geolocation data can also facilitate criminal activities such as stalking, robbery, and even kidnapping, as such data can easily pinpoint a user's current or future location.

Another study conducted by FTC on twelve health-related apps showed that sensitive health conditions such as pregnancy status, gender, and ovulation information were transmitted to 76 third parties including advertisement and analytics firms. Further, a recent study estimated that the information obtained from health records holds about fifty times more value than credit card information as such information can be easily used in identity theft and other fraudulent activities.

Furthermore, the privacy violations of the glasses-based wearables are widely acknowledged. One survey showed that the vast majority of surveyed individuals indicated that they would feel uncomfortable if a smart glass user records a video or snaps a photograph of them without their consent and considered it as a violation of their privacy. It is also worth mentioning that many entertainment businesses such as movie theaters and casinos have banned the use of such devices in their facilities due to privacy, and security, and copyright related concerns.

In a workplace environment, storing and transferring sensitive data using wearables could violate privacy laws such as The Health Insurance Portability and Accountability Act (HIPAA)[1] or the firm's Intellectual Property. One proposed solution from business leaders is to create new organizational rules and update the firm's network security infrastructure in order to detect and potentially control the data traffic from and to wearables. Another possibility is to consider implementing a mobile device management (MDM) system to manage what features are enabled or disabled on the wearable device and the smartphone. Furthermore, advanced security solutions could enable the analysis of data flows and could aid in identifying the type of device transmitting any data. For example, in platforms where wearable devices exist, a network administrator could be alerted when an out of network data communication takes place. Even if this technique may not be able to block the communication, detecting the transmission generated from the wearable device may be sufficient to inform a network administrator that an unauthorized device is being used on the network.

9.3 Psychological and Social Concerns

The twenty-first century is witnessing substantial technological changes which have given rise to new social and psychological implications.

Along with the growth that IoT and wearable technology have undergone in the past decade come new behavioral trends which may be indicative of social and psychological influences among users. Hence, it is never early to address and consider the potential adverse impacts of such effects and influences created by current and foreseeable trends.

For example, while it is widely accepted that smartwatches greatly enhance our connectivity, and allow us to attain regular tasks more conveniently, we have to bear in mind that such technology is becoming increasingly invasive. It is also important to consider the degree to which this could elevate stress levels and anxiety.

---

1 **The Health Insurance Portability and Accountability Act (HIPAA)** generally sets the US standards for protecting health information, which may consist of electronic and other forms of media containing identifiable information concerning an individual's past, current, or future physical or mental health that is generated or received by healthcare providers or employers.

Studies show that the average smartphone user unlocks their device 110 times a day, a number that would undoubtedly increase once wearables become more mainstream. Another study shows that the current generation spends about half of their time thinking about something other than what they are intended to be doing which affects their happiness and satisfaction levels. This is largely attributed to the device's distraction which diverts an individual from being immersed in a genuine experience such as being sincerely engaged in a conversation, appreciating a scenic view, or enjoying a good meal, to focusing on the continuous interruptions from a phone's text, app, or a social media notification.

Despite the many lifestyle benefits associated with IoT and wearable technology, we need to also consider their potentials to disrupt users' lives and the adverse effects they cause on their social and mental well-being, which will be the discussion topic of this section.

9.3.1 Psychological Concerns

While it is generally agreed that modern communication technologies are useful and helpful, and make the lives of people easier, they may undeniably make us restless, anxious, subject to frequent distractions, and always in need for constant entertainment. What is not helping is that the pace at which technology is progressing is so fast that our psychological processes are not keeping up.

In a book titled "iDisorder," the author hypothesizes that many technology users today could be diagnosed with what he calls an iDisorder. The author describes the psychological disorder as follows: "An iDisorder is where you exhibit signs and symptoms of a psychiatric disorder such as Obsessive–Compulsive Disorder (OCD), narcissism, addiction or even Attention Deficit/Hyperactivity Disorder (ADHD), which are manifested through your use, or overuse, of technology." A compulsive desire to check for text messages or emails, a desperate need to constantly update your Facebook status, or an obsessive addiction to iPhone games are all indications of iDisorder. There is no doubt that technology is affecting the way our brains function, whether or not these behavioral changes due to technology use are classified as a disorder, or a form of mental illness.

Smartphones have brought a groundbreaking level of convenience to people's lives. On the flipside, they make us accessible at any minute to colleagues, employers, friends, and relatives which may not always be ideal.

The way in which technology now is integrated within our lives means that it is becoming harder to achieve a work–life balance. The key challenge today is about attaining the willpower to withstand checking or responding to work-related emails while at home or on vacation. With more enterprises implementing wearable technology and other smart devices, new workplace culture would further lengthen and intensify the working day, which may turn the minor mental issues into real mental health problems.

For example, in the case of Swisscom Chief Executive Carsten Schloter's suicide, media reports suggest that he had become dangerously addicted to his smartphone. While this may represent an extreme outcome of the pressures resulted from being constantly connected, psychological studies confirm that smartphones are indeed introducing a new form of stress for users at home, work, and in social environments.

To prevent a new age of connected workplaces from further affecting the employee's mental health, employers will need to invest in tracking the extent of technology-related mental health issues, such as email addiction, that arise across the workplace. Employees need to be educated about the potential risks associated with using wearable technology, including the potential of being addicted to the device. This can be done through seminars and workshops offered at the workplace as a preventative measure as well as providing resources for employees that may have developed an addiction already.

A recent study shows that more than one third of children under the age of 2 use some form of mobile media. While another report confirms that as children age, this percentage dramatically increases with 95% of US teens (12–17) spending around nine hours online on average, while the average is about six hours per day for kids between the ages of 8 and 12.

Another study reports that the use of technology can modify the actual wiring of the brain. The time spent with technology alters the way teens' brains work. For instance, the study shows that while videogames may condition the brain to pay attention to several stimuli simultaneously, they may lead to distraction and reduced memory. Kids who constantly use search engines may excel at finding information, but are not good at remembering it. Moreover, the study reports that kids who use technology excessively may not have sufficient opportunities to use and develop their imaginative side or to read and deeply reflect about a given material.

Results of a recently published research show that higher levels of texting are correlated with poorer quality of sleep more likely because the study subjects felt obliged to respond to texts received during the night. Furthermore, excessive texting activity was correlated with elevated difficulties in stress management for those already experiencing some form of stress. It can certainly be predicted that the use of IoT-based personal assistants and wearable devices that enable hands-free operation and immediate access to unlimited information will likely promote the severity of the abovementioned issues.

9.3.2 Social Concerns

Although the majority of people enjoy a moderate use of social networking, a percentage of users have problems controlling the amount of time spent online. Many psychologists relate the urge to visit social media sites to addictive behaviors. Users who are addicted to social media often prefer online communication over face-to-face communication. They spend a disproportionate amount of time on their smartphones, tablets, or wearables because it enables them to control social interactions and avoid many of the possible uncertainties involved in direct face-to-face contact.

The social media addiction or overuse has intensified with the widespread availability of internet-enabled mobile phones and other handheld devices. Many users visit their social media accounts while walking down the street, attending a business meeting, or dining at a restaurant, taking advantage of the unconstrained connectivity. The emergence of wearables gives users even more accessibility to social media sites which would undoubtedly further intensify the addictive side effects. In fact, social science researchers and professionals warn that some users may become very dependent on online social media to the point of neglecting essential aspects of their off-line existence, such as their jobs, family, friendships, and health.

A recent study conducted on two groups of sixth graders revealed that the group who abstained from the use of any electronic devices for five days achieved better scores at picking up on emotions and nonverbal gestures of photographs of faces than the other group that used electronic devices. The increase in face-to-face interaction that the first group experienced resulted in them being more sensitive to facial expressions. The study concludes that the use of technology can affect a child's ability to empathize.

Another study found that kids who use videogames and online media for more than four hours a day do not have the same perception of well-being compared to those who used that technology for less than one hour. Many social experts agree that with less physical contact, kids will have difficulty developing social skills and emotional responses.

The increasing use of mobile devices and their wearable complements (i.e. smart watch) at events, restaurants, and other social venues has generated a substantial criticism as well.

In conclusion, it is clear that IoT and wearable technology will revolutionize the twenty-first century, and will continue to permeate our society. At the same time, it is very probable that, like earlier technological trends, these technologies will introduce unprecedented social and mental health issues. Hence, it would be smart to take a proactive approach to address the psychological and social impacts of these technologies. One way to do this is for researchers in the technical, social, and psychological science fields to collaborate in conducting research on the possible effects and impacts of these transformative technologies.

9.4 Safety Concerns

IoT and wearable technology are taking the concept of safety to a completely different level. Physical contact becomes optional in a connected world, and verbal, even gestural, commands can be used to operate and control devices, which give rise to new safety implications that have to be taken into consideration.

For example, IoT products can increase the risks of overheating, electric shocks, auditory hazards, etc. Moreover, controlling operations remotely means that such hazards are no longer limited by physical proximity. For example, an IoT-based oven that can be turned on and off remotely can become dangerous if a faulty command is received when no one is present to monitor it.

Regulations and standards that address the abovementioned hazards are being continuously created and revised. Smart appliances and home devices such as smart cooking pots, light bulbs, and thermostat are the frontline candidates in IoT safety. Self-driving cars, drones, and robotic assistants with the capability of injuring people or causing damage to properties and assets are also being considered. As IoT and wearable technology continue to evolve, it is important to stay informed about risks and safety standards and make sure products meet regulatory requirements.

9.5 Health Concerns

This section reports the major health concerns that come with using IoT and wearable technology raised by research studies from academia and health organizations, in addition to recommendations on ways to minimize such potential risks.

9.5.1 Electromagnetic Radiation and Specific Absorption Rate

With the emergence of IoT and wearable devices, more reports have surfaced discussing the potential hazards of being around continuous RF electromagnetic radiation. This is the same type of energy radiated by cell phones, tablets, and laptops, and studies have been around for some time which allow for safety comparisons between old and new technologies.

Most wearable products utilize BLE technology, which emits lower levels of RF energy compared to cell phones and other WiFi-enabled devices. While it is confirmed that exposure to electromagnetic radiation from wireless devices induces heating in the area where they are held, experts argue that wearables are much safer since they radiate lower energy levels than cellular and WiFi-based devices.

However, new research is beginning to reveal inconvenient findings about the nonthermal effects caused by radiation, and hundreds of experts are calling on governments to issue precautionary statements and adopt regulations that protect consumers from these risks.

Electromagnetic waves in the radiofrequency range (3 KHz–300 GHz) are what enable transmission of wireless telecommunications including cellular networks, television, and radio broadcasting, in addition to IoT and wearable devices. These waves are radiated by antennas which are designed with different sizes and shapes to allow for different frequencies of operation and radiation patterns which specify the direction and distribution of the radiated energy.

It should be noted that the human body absorbs some of the electromagnetic energy radiated from wireless devices. The amount of the absorbed energy is calculated using a measure called "Specific Absorption Rate (SAR)," which is expressed in Watts per Kilogram of body weight. The Federal Communications Commission (FCC) mandates that every wireless device sold in the United States must be tested and verified to have a SAR less than 1.6 Watts/Kilogram (W/Kg) before it can go on sale. Canada's regulations also mandate a 1.6 W/Kg limit, while the European Union and Australia require 2.0 W/Kg.

It is worth noting that SAR measurements have received a great deal of criticism in the past few years, and many organizations consider SAR as an unreliable measure of whether or not a wireless device is safe. A mobile phone, for example, may have a SAR of 0.9 W/Kg, but that may not be any safer than a device with 1.2 W/Kg. A phone's SAR value, for example, can vary widely during a call as the device alternate between transmission channels and as the distance from a cellular tower is increased.

The FCC Guide "Specific Absorption Rate (SAR) For Cell Phones: What It Means for You" states: "ALL cell phones must meet the FCC's RF exposure standard, which is set at a level well below that at which laboratory testing indicates, and medical and biological experts generally agree, adverse health effects could occur. For users who are concerned with the adequacy of this standard or who otherwise wish to further reduce their exposure, the most effective mean to reduce exposure is to hold the cell phone away from the head or body and to use a speakerphone or hands-free accessory. These measures will generally have much more impact on RF energy absorption than the small difference in SAR between individual cell phones, which, in any event, is an unreliable comparison of RF exposure to consumers, given the variables of individual use." Moreover, SAR limits set by the FCC do not take into consideration that the human body is also sensitive to the power amplitudes and frequencies responsible for the microwave hearing effect, also known as the Frey effect that occurs with exposures of $400\,\mu W/cm^2$, which is well below FCC's SAR limits.

The Frey effect, which was first reported by individuals working in the vicinity of radar transmitters during World War II, consists of audible clicking and buzzing induced by pulsed radio frequencies. These audible clicks are generated inside the head without the need of any receiving antenna. The cause is thought to be the thermoelastic expansion of parts of the middle and inner ear.

It should be noted that many IoT and wearable devices use BLE technology, which emits much lower power than classic Bluetooth, and significantly less than cell phones. In fact, in some cases, ultra-low power devices are not required by the FCC to be tested for SAR as opposed to cell phones and laptops which must pass a rigorous testing. However, most IoT devices and some wearables do not limit their on-board wireless capability to Bluetooth and may use WiFi too, which is comparable to cell phones in terms of electromagnetic energy radiation.

As mentioned previously, antennas are what enable wireless communication in electronics. In wearables, antennas are required to be compact, lightweight, and mechanically robust. They are also preferred to be flexible with a low profile (thin); yet, they must express high efficiency and desirable radiation characteristics.

To minimize SAR in wearable devices, antennas are preferred to have a uni-directional (hemi-spherical) radiation pattern, radiating away from the user's body to reduce the user's exposure to electromagnetic radiation. However, antennas that offer such characteristics, like microstrips, suffer from a relatively narrow bandwidth which is a function of the platform (substrate) thickness. Thus, the majority of handheld electronics designers choose printed monopole\dipole antennas which offer a simple, thin, compact, and cost-effective solution, but also exhibit an omni-directional radiation pattern (i.e. radiates in all directions including the user's body) (Figure 9.1).

Many antenna designs have been proposed in the literature to resolve the SAR and thickness trade-off. One design proposed by the author of this book features a low profile printed monopole antenna integrated with a compact artificial magnetic conductor (AMC) ground plane which is utilized to provide the desired uni-directional radiation pattern while keeping a thin antenna profile. The proposed design offers a 65% reduction in SAR while maintaining a relatively large bandwidth and a compact design.

Lastly, it should be noted that there are no enough studies have been conducted on the adverse effects of 5G-enabled devices which use much higher frequencies than the conventional wireless technologies.

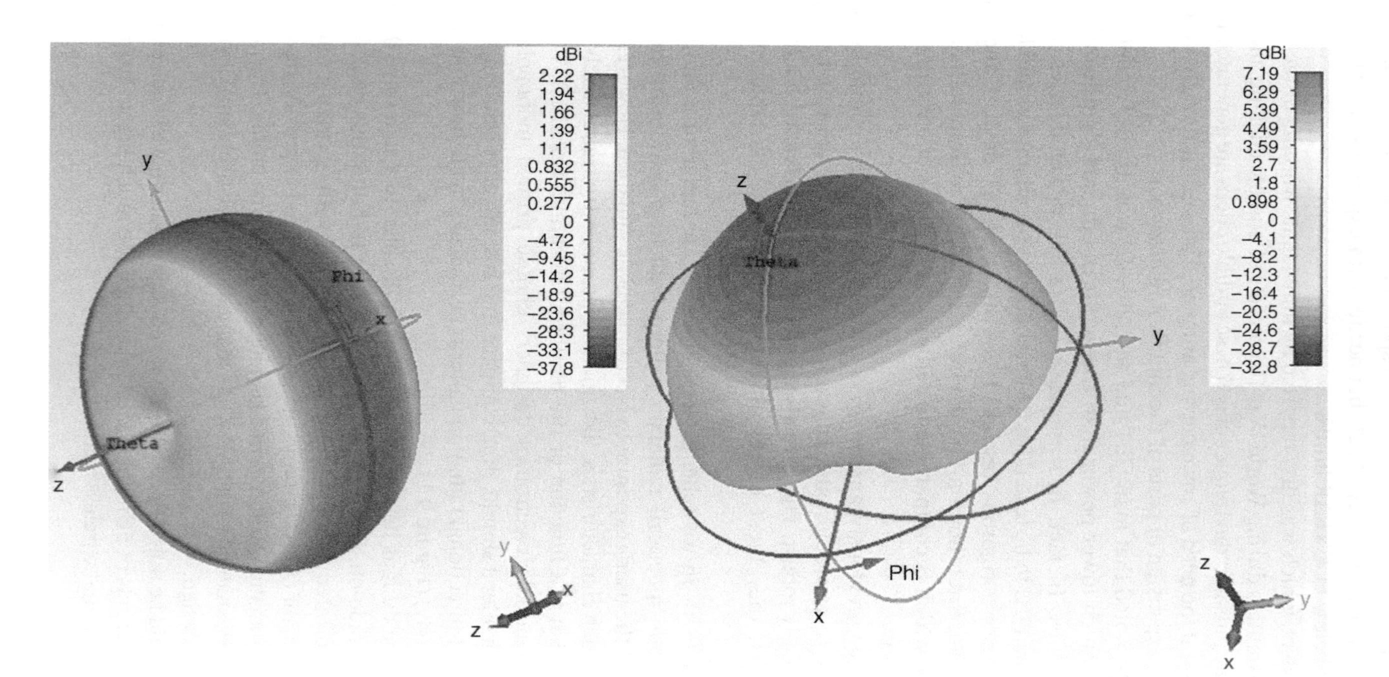

Figure 9.1 Omni-directional radiation pattern (left), and semi-directional (hemi-spherical) radiation pattern (right).

9.5.2 Diseases and Effects

9.5.2.1 Cancer

Exposure to ionizing electromagnetic radiation such as X and Gamma rays is known to increase the risk of cancer. However, the numerous studies that have examined the potential effects of nonionizing RF radiation on health from microwave ovens, cell phones, radars, and other wireless transmitting sources confirm that there is no consistent evidence that it could increase the risk of cancer.

The only recognized biological effect of RF energy at this time is hyperthermia, which is defined as an increase in body temperature that occurs when the body generates or absorbs more heat than it dissipates. The ability of microwave ovens to heat water particles in food is one example of the effect of RF energy.

However, some studies report that even radiation from low energy devices could be problematic due to the blood–brain barrier effect. According to these studies, exposure to low energy radiation could trigger the opening of the blood–brain barrier which allows toxins in the blood stream to penetrate the brain tissues even with low exposure to electromagnetic radiation.

The International Agency for Research on Cancer (IARC), a WHO entity, classifies the use of cell phones as "possibly carcinogenic," based on limited evidence from human and rodent subjects and inconsistent results from mechanistic studies. While the National Institute of Environmental Health Sciences (NIEHS) reported that the current scientific evidence that associates cell phone use with any negative health effects is not conclusive, and NIEHS did state that more research is needed.

The U.S. Food and Drug Administration (FDA), on the other hand, stated that human epidemiologic studies claiming biological changes linked to RF energy exposure have failed to be replicated, while the U.S. Centers for Disease Control and Prevention (CDC) reported that no scientific evidence conclusively answers whether cell phone use leads to cancer.

The Federal Communications Commission (FCC) confirmed that there is no scientific evidence to establish a link between radiation from wireless devices and cancer. The European Commission Scientific Committee echoes the FCC's report and also concluded that epidemiologic studies do not indicate an elevated risk for other malignant diseases, including childhood cancer.

Other European studies concluded that talking on a mobile phone for extended periods could triple the risk of brain cancer. In contrast, a study published in the British Medical Journal found that there was no proof of increased cancer. However, it is worth noting that the researchers behind the reported study acknowledged that a small to moderate increase in cancer risk must not be ruled out, especially among heavy cell phone users.

9.5.2.2 Fertility

Several epidemiological studies have found reductions in sperm quantity, motility, and viability in male subjects using mobile phones for more than a few hours per day. The production of reactive oxygen species (ROS) which can potentially cause damage to cell membranes and DNA was a common finding associated with the reported effects. Another study echoes these findings which found that ejaculated semen from healthy donors exhibited reduced viability and motility, and an elevated ROS levels after one hour of exposure to a cell phone in talk mode.

A more recent study found that exposing ejaculated sperms to WiFi radiation from a laptop for four hours had led to reduced sperm motility and increased DNA fragmentation when compared to samples exposed an identical laptop with the WiFi capability turned off. It should be noted that to date, there has been no strong evidence that cell phone radiation affects female fertility.

While most wearables utilize low energy communication schemes, it is the hub such devices need to forward, store, and process information (i.e. cell phone and tablet) that emit higher electromagnetic energy which could trigger the aforementioned effects.

9.5.2.3 Vision and Sleep Disorders

A plethora of studies have found that the blue light emitted from electronics such as cell phones, tablets, and laptops reduces the production of the sleep-regulating hormone: melatonin. Research findings are clear that people who excessively use their laptops and smartphones are more prone to experience symptoms of insomnia.

Moreover, some studies report that blue light may cause retinal damage which could lead to macular degeneration, a known cause of blindness. Although this damage is attributed to direct exposure which is a far greater level of exposure than a user would get from the display of a handheld or a wearable device, researchers have just begun to understand the effects of blue light exposure on vision. Hence, caution should be used especially with emerging head-mounted wearables that use projectors and augmented and virtual reality technologies.

Furthermore, screens on mobile and handheld electronics tend to be smaller than computer displays, which likely force users to squint and strain their eyes while reading messages or navigating an app. According to The Vision Council of America, more than 70% of Americans are not aware or are in denial that they are prone to digital eye strain, which is a temporary discomfort that follows a few hours of digital device (with a display) use.

9.5.2.4 Pain and Discomfort

Some studies correlate joint pain and inflammation to the unnatural rapid movement of hands during the use of handheld device and wearables. Back and neck pain is also common with prolonged cell phone use, especially if it is held between the neck and shoulders as the user multitasks. One study reported that long hours of cell phone use cause users to arch their bodies in an unnatural posture which can lead to back and neck pain. The abovementioned scenarios are also applicable to wearables that may influence unnatural postures and orientation changes.

Additionally, the findings of some studies suggest that excessive electromagnetic radiation could also stimulate the production of adrenalin and cortisol which may cause headaches, cardiac arrhythmia, high blood pressure, and tremors. However, these effects are generally triggered by high power radiators such as cellular base stations and are less likely to be caused by low power radiation from wearables.

9.5.2.5 Other Risks

One study on fetal development reports that fetuses exposed to electromagnetic radiation from their pregnant mothers' phones can trigger childhood behaviors such as hyperactivity, reduced short-term memory, and ADHD.

Although contradicted by other studies, the secretion of cortisol, which is often referred to as the stress hormone, has been shown in one study to be affected by RF exposure. It is assumed that RF radiation may serve as a stressor evident from the elevated cortisol concentrations reported in a number of studies involving animals and humans.

Effects of RF energy localized to the head have been studied repeatedly. Results suggest that RF energy exposure on blood flow in the brain has no hazardous effects. Some studies suggested that RF energy might affect the metabolism of glucose, but follow-up studies that examined glucose metabolism inside the brain after cell phone exposure showed inconsistent findings.

Lastly, when it comes to hygiene, the continuous touching of handheld and wearable electronics can foster germs on the device. The greasy residue a user's hand leaves on the device after a day of use can accumulate more disease-causing germs than those found on a toilet seat. For example, one study found that 92% of the 390 smart phones sampled had bacteria on them, 82% of the users' hands had bacteria, and 16% of smart phones and corresponding hands had *Escherichia Coli.* fecal matter. This would still apply to wearable devices and other gadgets even if it's to a different extent.

9.5.3 Recommendations

Unlike cell phones, wearables are intended to be in close contact with the user's body and for far more extended durations. While there are established data on the effects of using cell phones, not enough time has elapsed for research to agree on the health risks and exact impact of wearable devices; hence, it is recommended to use caution and common sense.

More people nowadays are abandoning their landline service and are exclusively relying on their mobile phones. Also, driving laws are regulating the use of mobile devices which has resulted in higher use of Bluetooth ear phones while driving. Therefore, it is advisable to use the built-in hands-free feature (available in most modern cars), wired earphones, or smart home units to minimize low energy radiation effects whenever possible. It is also recommended to avoid positioning any wearable or mobile device in close proximity to the reproductive organs for extended periods of time. These devices should be kept out of pants' pockets. For female users, it is recommended to not place a wearable device within 15 cm of the breast.

The use of mobile phones and cellular-based wearables should be limited to areas with excellent reception. The weaker the reception, the more power the cellular-based IoT or wearable device will have to emit, which means higher electromagnetic energy deposition in the user's body. Obviously, children would be at higher risk from radiation due to their thinner skulls and still developing nervous systems. WHO has stated that the farther away a wireless device is from the user's head, the less harmful it would be.

Lastly, wearing head-mounted wearable devices while asleep and placing IoT and wearable devices on the nightstand next to the head or under the pillow should be completely avoided.

9.6 Regulations

IoT and wearable devices are still largely unregulated, with no specific laws or regulations governing how data from these devices are collected or used by parties other than the user. In light of the security and privacy concerns associated with these technologies, several bodies have called on the US government, including federal agencies and Congress, to undertake a more active role in coordinating regulation and standards.

Several U.S. agencies including the Department of Commerce, Department of Defense (DoD), and Department of Justice have some form of IoT regulation, but this has led to overlapping responsibilities which created bureaucratic challenges. The need for one inclusionary authority is obvious.

Moreover, experts argue that it is difficult for the industry to develop industry-wide standards due to the fact that the privacy and security of IoT and wearable technology fall upon several actors, including manufacturers, network providers, software developers, and other third parties.

In response to such challenges, the FTC, National Telecommunications and Information Administration (NTIA), Food and Drug Administration (FDA), the National Institute of Standards and Technology (NIST), and representatives from the U.S. Senate and House of Representatives took the initiative to address IoT privacy and security concerns by conducting several meeting between 2016 and 2018. Moreover, a public–private sector working group assembled by the NTIA finalized a guidance document addressing how manufacturers should convey information to consumers concerning security updates for IoT and wearable devices.

Moreover, in 2017, a bipartisan bill aiming at improving the cybersecurity of IoT devices supplied to the U.S. government was introduced by the senate. The bill comprised several provisions, including requirements that supplying vendors guarantee that devices are governed by industry standard protocols, are not based on hard-coded passwords, and do not have any known security vulnerabilities.

The FTC has viewed IoT security as a priority. They provided a set of recommendations for best practices businesses can enforce in order to protect the privacy and security of consumers. In fact, the FTC not only supported giving notice to users about what data are being collected, but also providing them a choice of how their data are to be collected and shared.

Regulation proponents believe that standards applied to every product would help protecting the security of users if the government is to regulate IoT and wearable technology. Others, however, expressed concerns if IoT is to be regulated by the government. These concerns include the elimination of smaller businesses which would compromise market competition and consumer choice, innovation impediment due to bureaucracy, and the lack of government expertise to effectively regulate these technologies.

In addition to calling on the federal government to regulate the IoT and wearable technology, experts have also urged the private sector to engage in self-regulation.

One action was that some of the major industry players such as Google and Sprint backing the British chip designer ARM's security framework referred to as the Platform Security Architecture (PSA). The objective was to create a common industry framework for every IoT product. According to ARM, about 100 billion devices are already using their designs, and this number is expected to double by 2021. PSA consists of "threat models, security analyses, hardware and firmware architecture specifications, and an open source firmware reference implementation," which, collectively, provide a foundation for security to be consistently integrated into the devices at the hardware and firmware levels.

On the other hand, the European Union (EU) has also taken action concerning IoT and wearable devices, such as passing the General Data Protection Regulation

(GDPR) and the ePrivacy Regulation. Another action was by the European Parliament (EP) where they recommended passing the Privacy Impact Assessments concerning the RFID applications privacy and data protection framework to wearables. The EP also recommended deleting raw data once processed, and immediately informing the user once a data compromise risk is detected.

These regulations exemplify the different approaches pursued by the EU and the United States. The EU uses a holistic approach, providing European citizens with certain privacy rights across all platforms and sectors. The United States, on the other hand, has an agglomeration of privacy laws specific to different industries. Furthermore, the U.S law typically balances privacy rights and interests against freedom of expression, which is driven by the First Amendment, while the EU firmly asserts that privacy is a fundamental right, and the way personal data are used by third parties should be governed by regulation, controls, and transparency, which requires government supervision.

Further Reading

Bergman, N. and Rouse, J. (2013). *Hacking Exposed Mobile: Security Secrets & Solutions*, vol. 1. New York: McGraw-Hill.

Bowman, J.D., Kelsh, M.A., and Kaune, W.T. (1998). *Manual for Measuring Occupational Electric and Magnetic Field Exposires*. Washington, DC: U.S. Department of Health and Human Services.

Buenaflor, C. and Kim, H.-C. (2013). Six human factors to acceptability of wearable computers. *International Journal of Multimedia and Ubiquitous Engineering* 8 (3): 295–300.

Cellular Phone Towers (2015). Center for health, environment & justice. FactPack - PUB 129.

Clayton, R.B., Leshner, G., and Almond, A. (2015). The extended iSelf: the impact of iPhone separation on cognition, emotion, and physiology. *Journal of Computer-Mediated Communication* https://doi.org/10.1111/jcc4.12109.

Common Sense Media (2011). Zero to eight: a common sense media research study children's media use in America. *FALL*.

Dart, P., Cordes, K., Elliott, A. et al. (2013) Biological and Health Effects of Microwave Radio Frequency Transmissions: A Review of the Research Literature. *A Report to the Staff and Directors of the Eugene Water and Electric Board*.

Desai, N.R., Kesari, K.K., and Agarwa, A. (2009). Pathophysiology of cell phone radiation: oxidative stress and carcinogenesis with focus on male reproductive system. *Reproductive Biology and Endocrinology* 7: 114.

European Commission (2011). Privacy and data protection impact assessment framework for RFID applications.

Federal Communications Commission (2016) SAR for cell phones: what it means for you, consumer guide. Federal Communications Commission.

Frey, A.H. (1962). Human auditory system response to modulated electromagnetic energy. *Journal of Applied Physiology* 17 (4): 689–692.

FTC Staff Report (2015). Internet of Things, Privacy & Security in a Connected World, FTC Staff Report.

Goh, J.P.L. (2015). Privacy, security, and wearable technology. *Landslides*, ABA Section of Intellectual Property Law 8 (2): 493–498.

Goldsworthy, A. (2012) The biological effects of weak electromagnetic fields problems and solutions.

Hardell, L., Carlberg, M., Söderqvist, F., and Mild, K.H. (2007). Long-term use of cellular phones and brain tumours: increased risk associated with use for more than 10 years. *Occupational and Environmental Medicine* 64 (9): 626–632.

Robert P. Hartwig, Claire Wilkinson, *Cyber Risk:Threat and Opportunity*, Insurance Information Institute, New York, 2015.

Hartwig, V., Giovannetti, G., Vanello, N. et al. (2009). Biological effects and safety in magnetic resonance imaging: a review. *International Journal of Environmental Research and Public Health* 6 (6): 1778–1798.

Harvard Health Letters (2012). Blue light has a dark side, what is blue light? The effect blue light has on your sleep and more. *Harvard University, Harvard Health Letters* (May 2012). http://www.health.harvard.edu/staying-healthy/blue-light-has-a-dark-side.

Irish Council for Bioethics (2009). *Biometrics: Enhancing Security or Invading Privacy?* Dublin, Ireland: The Irish Council for Bioethics.

Kahina, C. (2015). Security issues in wireless sensor networks: attacks and countermeasures. *Proceedings of the World Congress on Engineering 2015 Vol I, WCE 2015*, London, UK (1–3 July 2015).

Ke, W.C. and Singh, M.M. (2016). Wearable technology devices security and privacy vulnerability analysis. *International Journal of Network Security & Its Applications (IJNSA)* 8 (3).

Kirtley, J. and Memmel, S. (2018). Rewriting the "book of the machine": regulatory and liability issues for the internet of things. *Minnesota Journal of Law, Science & Technology* 19 (2): 455–513.

Knight, J.F. and Baber, C. (2005). A tool to assess the comfort of wearable computers - human factors. *The Journal of the Human Factors and Ergonomics Society* 47 (1): 77–91.

Kuss, D.J. and Griffiths, M.D. (2011). Online social networking and addiction—a review of the psychological literature. *International Journal of Environmental Research and Public Health* 8 (9): 3528–3552.

Lanhart, A., Purcell, K., and Smith, A. (2010). *Social Media and Mobile Internet Use Among Teens and Young Adults*. Washington, DC: Pew Research Center.

Murphy, D.L., Timpano, K.R., Wheaton, M.G. et al. (2010). Obsessive-compulsive disorder and its related disorders: a reappraisal of obsessive-compulsive spectrum concepts. *Dialogues in Clinical Neuroscience* 12 (2): 131–148.

Peart, K.N. (2012). Cell phone use in pregnancy may cause behavioral disorders in offspring. *Yale News (Article)* (15 March 2012).

Raad, H., Abbosh, A.I., Al-Rizzo, H.M., and Rucker, D.G. (2013). Flexible and compact AMC based antenna for telemedicine applications. *IEEE Transactions on Antennas and Propagation* 61 (2): 524–531.

Shih, P.C., Han, K., Poole, E.S., and Rosson, M.B. (2015) Use and adoption challenges of wearable activity trackers. *iConference*, Newport Beach, CA, USA, pp. 1–12.

Shokri, S., Soltani, A., Kazemi, M., and Sardari, D. (2015 Summer;). Effects of Wi-Fi (2.45 GHz) exposure on apoptosis, sperm parameters and testicular histomorphometry in rats: a time course study. *Cell Journal* 17 (2): 322–331.

Subrahmanyam, K., Kraut, R.E., Greenfield, P.M., and Gros, E.F. (2000). The impact of home, computer use on children's activities and development. *Children and Computer Technology* 10 (2): 123–144– Fall/Winter.

The Vision Council (2015) Digital Eye Strain Report.

Uhls, Y.T., Michikyanb, M., Morrisc, J. et al. (2014). Five days at outdoor education camp without screens improves preteen skills with nonverbal emotion cues. *Computers in Human Behavior* 39: 387–392.

Vermesan, O. and Friess, P. (2013). *Internet of Things: Converging Technologies for Smart Environments and Integrated Ecosystems*. Aalborg, Denmark: River Publishers.

Volkow, N.D., Tomasi, D., Wang, G.-J. et al. (2011). Effects of cell phone radiofrequency signal exposure on brain glucose metabolism. *JAMA* 305 (8): 808–813.

Warren, S.D. and Brandeis, L.D. (1890). The right to privacy. *Harvard Law Review* 4 (5): 193–220.

Yano, K., Ara, K., Watanabe, J. et al. (2015). Measuring happiness using wearable technology, technology for boosting productivity in knowledge work and service businesses. *Hitachi Review* 64 (8): 517–524.

10

Detailed Product Design and Development: From Idea to Finished Product

Scott Tattersall, Mustafa Kamoona, and Haider Raad

10.1 Introduction

Now that we have learned the ins and outs of IoT and Wearable Technology, it is time to apply the knowledge learned in the previous chapters to design and develop two complete practical products. This chapter will take the reader, step by step, from concept and engineering requirements through bread-boarding, microcontroller coding, PCB design, PCB printing, soldering, and surface mount considerations all the way to a finished product.

The first product is an IoT connected device aimed at helping vineyard owners to remotely monitor the moisture level of the vineyard soil as well as ambient conditions such as temperature, humidity, and light levels at various points during the day. The second product is a wearable solution that can reliably detect an elderly person's accidental fall, and contact emergency for help. While some of the steps discussed in the design and consideration chapter were not used/needed in one product, it is hoped that the processes and methodologies used in the two products complement each other collectively.

10.2 Product I (IoT): Vineyard Monitor

The product that will be designed and developed in this section is a device that helps vineyard owners monitor the moisture level of the vineyard soil as well as ambient conditions such as temperature, humidity, and light levels at various points during the day. This will help vineyard owners track the conditions for their vines and understand the best and most efficient watering strategy, which is a key consideration from a quality perspective of grape/wine production and also from an environmental impact perspective.

Fundamentals of IoT and Wearable Technology Design, First Edition. Haider Raad.

Published 2021 by John Wiley & Sons, Inc.

The functionality choices here allow for a good demonstration of how to design hardware with multiple components and how to integrate the outputs from these components together into one neat package.

10.2.1 Product Requirements and Design Considerations

As mentioned in Chapter 6, when pursuing a new product development, it is essential to define the requirements and any necessary design considerations.

The high-level product requirements that usually come from the customer are broken down into measurable and verifiable components. Depending on the application and its complexity, this list can range from a few requirements in relatively simple projects to numerous requirements in complex industrial solutions. The vineyard monitor system requirements and design considerations are listed below:

- Accurate data on moisture levels
- Accurate temperature and humidity measurements
- A relative measure of light level throughout the day
- Must work on a battery since the device will be used outdoors
- Must be small and easy to manage
- Must have a way to send information to the cloud for analysis and alerting
- Must be inexpensive since ideally a vineyard owner would deploy a number of these devices to monitor the different conditions at various parts of the vineyard
- It must be possible to weatherproof this device since it will be outdoors, so ideally there would be no exposed electronic components, or external antenna that would need to be connected manually
- Since the device must run from a battery, it is important that its components should have exceptionally low power consumption and/or have a sleep mode that can be used when the component is not needed

10.2.2 Communication Network/Technology Selection

Based on the above requirements, one potential communication network to receive message data from the device is Sigfox. As shown in Chapter 5, Sigfox is one of the leading IoT service providers with a global network that allows billions of devices to connect to the Internet. Sigfox devices can send a small amount of data (12-byte message packets, which is perfect for the message data this device will be sending) a very long distance (up to 30 miles), with extremely low energy usage. Moreover, the Sigfox message costs are comparatively low, and hardware costs for Sigfox chips are low also, making it an attractive choice of communications technology. The following diagram shows the end-to-end building blocks of the solution, catered to the use case described in this section (Figure 10.1):

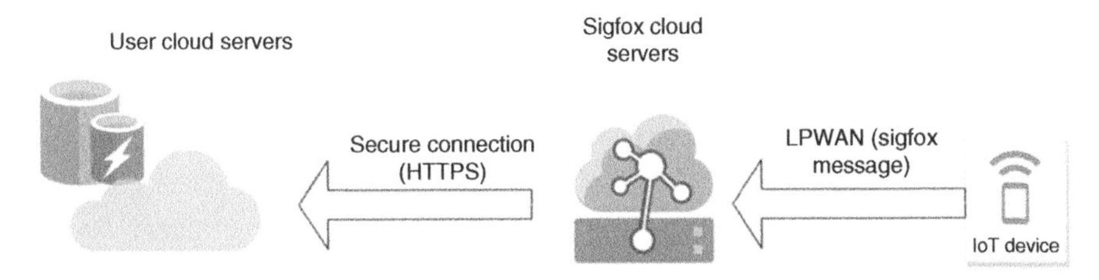

Figure 10.1 Vineyard monitor system block diagram.

10.2.3 Hardware Selection and Breadboarding

As all good electronics projects usually start, here we begin with Breadboarding out a conceptual circuit. A "Breadboard" is a plastic board with conductive lines underneath, and holes for fitting wires or plated-through-hole (PTH) components, which allows the developer to quickly connect components together and electrically test their operation before designing a PCB schematic.

Breadboarding involves working out functions you want your device to have and to achieve a proof of concept. For this project, we want the device to:

1) Consume low power. One reasonable choice of a microcontroller unit would be the ATMEGA328P, which is the same chip that an Arduino Pro Mini uses.
2) Have the ability to send messages over the Sigfox network. We chose Wisol SFMR10 for the following reasons:
   a) It is low power and has a small form factor
   b) It has versions for multiple Sigfox coverage regions
   c) There is a development kit (DevKit) available for it, which is very useful for breadboarding and prototyping

   We will also need to attach the Wisol chip with an antenna in order for it to be able to wirelessly send the messages to the Sigfox base stations. There are many antenna options available for the Sigfox network bandwidth, from PCB trace antennas, onboard antennas, and external antennas. For the small form factor we are considering, we will choose an onboard antenna from Proant (antenna shown in Figure 10.2). This antenna has a gain of 1.7 db and 2.4 dBi depending on the transmitting frequency (which is different for different Sigfox geographic zones).
3) Have three basic sensors:
   a) Temperature and humidity. We will consider using the DS18B20
   b) Light level. We will use a Light Dependent Resistor "LDR"
   c) Soil moisture sensor, one option could be the SEN-13322
4) Have a form of user interaction: The device may also need some way for the user to interact (reset/send message/self-test, etc.), but to avoid any mechanical components, we will use a magnetic sensor instead of a button.

Figure 10.2 Proant 868/915 antenna. *Source:* Photo courtesy of Proant.

10.2.3.1 Breadboarding Example

Below is an example of the general process of how to breadboard a single component, in this case, it will be the DS18B20 sensor that we will use in this project:

1) Read the datasheet [1] for the component you want to connect. From this, you may find that the component needs additional hardware, such as a smoothing capacitor or a pull-up or pull-down resistor in order for it to work correctly. Oftentimes, the datasheet will include a handy example schematic of how the device should be used, which will show you if any necessary components are needed. In this case, the datasheet includes the following example schematic (Figure 10.3):

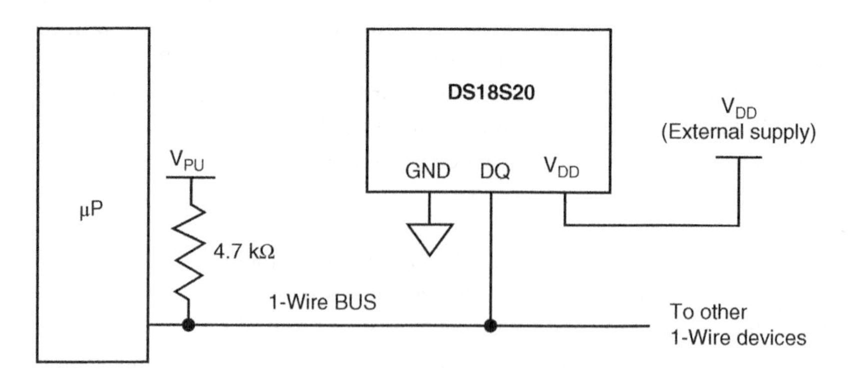

Figure 10.3 Sample schematic from DS18B20 datasheet. *Source:* [1]. © 2019 Maxim Integrated Products, Inc.

2) Next, we take an existing Arduino microcontroller, such as an Uno, or a Pro Mini if the component is 3.3 V only, and hook the component up to the relevant Arduino pins using a breadboard. Here, we will use an Arduino Uno, since the DS18B20 is able to run at 5 V and so our connections through the breadboard would look like this (Figure 10.4):

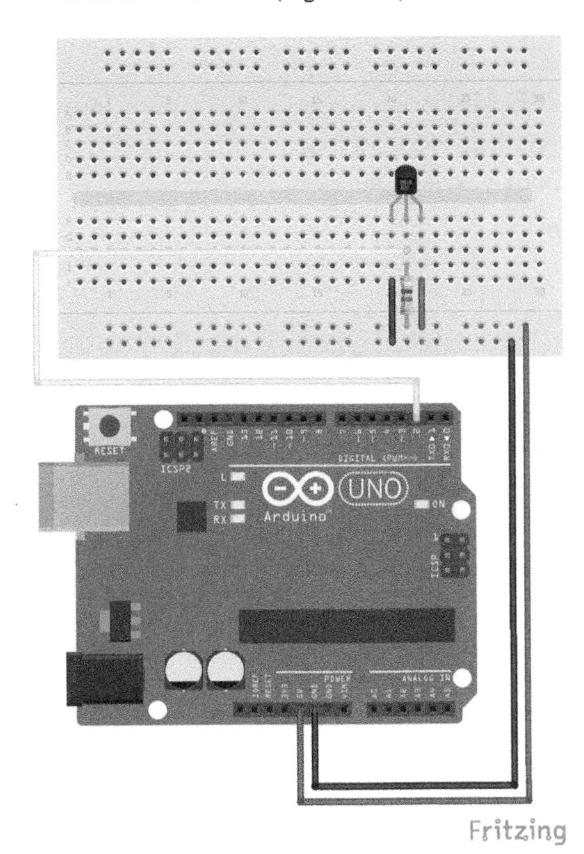

Figure 10.4 Breadboarding example showing DS18B20 connections.

3) Now we just need to write some Arduino code to read the temperature data at the right Arduino pin. Here, we may need to reference some external libraries; in this case for the temperature sensor, we will need to use the DallasTemperature library. To do this, simply select "Include Library" from the sketch menu and find the "DallasTemperature" by using the search box, and then click install. Once available, some sample code for printing out temperature data from this component, hooked up to the Arduino on Pin 2, would look like this:

```
// Include the libraries
#include <OneWire.h>
#include <DallasTemperature.h>

// set pin on Arduino receiving temperature data
#define TMP_PIN 2

// setup OneWire
OneWire temp_sensor(TMP_PIN);

// Pass reference to Dallas Temperature.
DallasTemperature sensors(&temp_sensor);

// Arduino Setup code
void setup(void)
{
   // start serial
   Serial.begin(9600);
   Serial.println("DS18B20 DEMO starting");
   // Start up the Dallas Temperature library
   sensors.begin();
}

// main Arduino loop code
void loop(void)
{
   Serial.print(" Requesting temperatures...");
   sensors.requestTemperatures(); // get readings

   Serial.print("Current tmp: ");
   Serial.print(sensors.getTempCByIndex(0));

   delay(1000); // wait 1 second before continuing the
main loop
}
```

The same kind of process can be used for all the other components we want to interface with as part of the project (for example, the light sensor and the soil moisture sensor). The result of this process should be a functional (if often somewhat messy!) set of components on a breadboard, in this case (Figure 10.5):

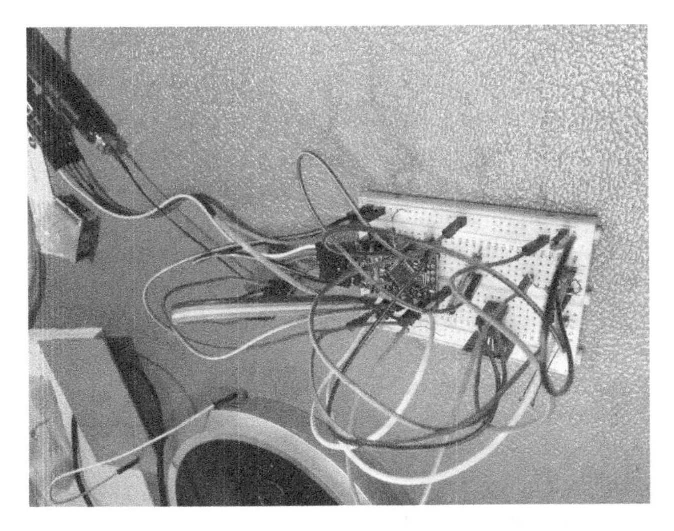

Figure 10.5 Initial working but untidy attempt at breadboarding.

However, it is recommended to spend a little extra time putting together a neater version using special breadboard "jumper" wires when you have everything working (see below) (Figure 10.6):

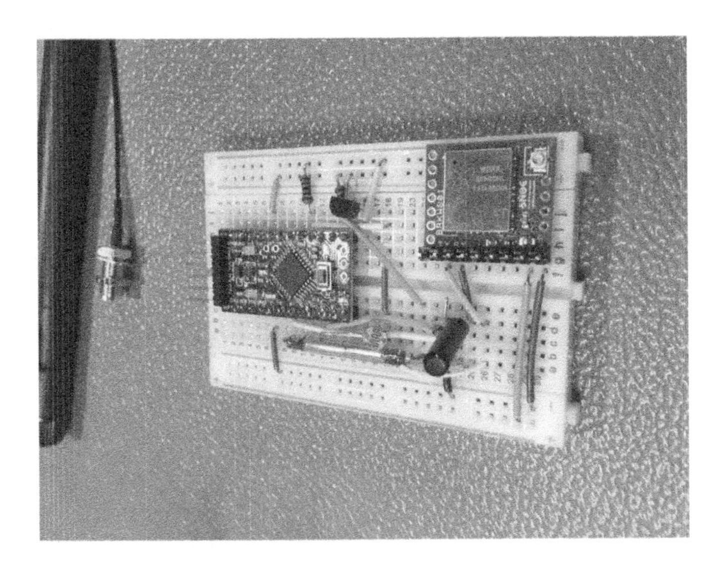

Figure 10.6 Tidier breadboard using jumper wires.

10.2.4 Prototyping

Once you are happy that your breadboarded solution is functional and working reliably, it is time to start thinking about designing the schematic from your breadboard. But before getting stuck into custom designing a PCB circuit for your device, it is very worthwhile trying to nail down a smaller, neater prototype circuit. Here, we opt for a stripboard version of this circuit.

Stripboard is a board consisting of a regular set of holes all connected along a row with copper (but each row is distinct from each other row). In this way, it is easy to solder components together in a layout that more closely matches how you want your end product to look. Also, the stripboard can be cut to size so you get a good idea of the necessary dimensions of your final product.

The end result should be a much neater and tighter version of the circuit, which is very useful in helping trim down the final PCB design (remember as a rule of thumb: the larger the PCB the higher the cost). It also gives you a good idea about what sort of housing may be required for your product.

10.2.4.1 Fritzing

Fritzing is a great piece of software for laying out Stripboard or Veroboard circuits, allowing you to completely produce a virtual circuit, which you can then simply copy on your stripboard. There are good tutorials on Fritzing, for example, see [2].

The prototype circuit for this project looks like this in Fritzing (Figure 10.7):

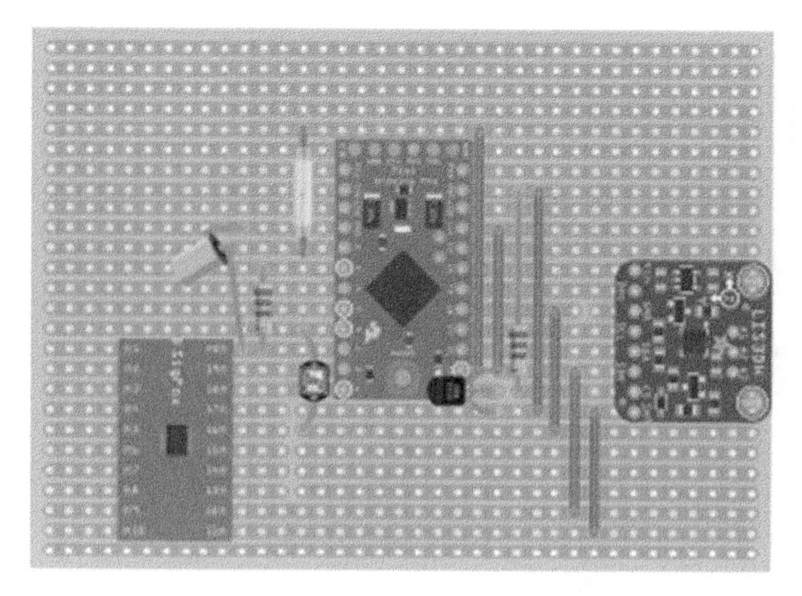

Figure 10.7 Frizing layout of stripboard.

Which leads to this actual (working) circuit (note how close the real board is to the Fritzing output) (Figure 10.8):

Figure 10.8 Final stripboard prototype.

10.2.5 Power Consumption

This is a surprisingly difficult but particularly important part of the process to complete. For this device, we want it to be ultra-low power so that it can work from a small battery (a 900 mAh CR2 battery was chosen in the end) for as long as possible. This means making sure that the Quiescent current (the constant current draw) is as small as possible, down into the low μA range, while accounting for the occasional higher current draw during message sending.

Quiescent Current can be defined as the amount of current used by an IC when in a Quiescent state. The Quiescent state being any period when the IC is in either a no load or non-switching condition yet is still enabled.

While there are a number of circuits and methods of assessing the current requirements of a circuit, most do not have a good resolution for the very low end and manual mechanisms (such as an ammeters connected across the power supply lines) are cumbersome to use and also only give snapshots of the current usage at a given time, and in some cases, do not react fast enough for any reliable measurement.

One solution to this problem is to use a dedicated low power measurement module, for example, the Power Profiling Kit from Nordic Semiconductor [3]. It is not too expensive (at around $100 for both the Power Profiler Kit (PPK) and the baseboard) and it works very well.

It produces both a constant view of the power consumption down to a very low resolution (<1 μA) and a running average for a time window (which is exactly what we need for our battery life calculations). Here is a sample output showing the Quiescent current draw as well as the power spike when a message is sent (due to the higher power requirement of the Wisol chip and RF communications in general) (Figure 10.9):

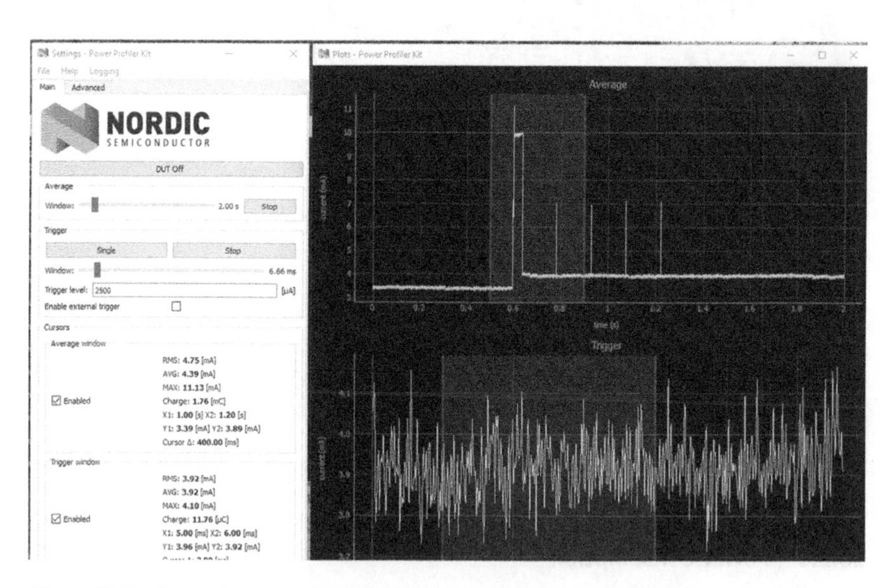

Figure 10.9 Example screenshot of power profiler. *Source:* nordicsemi.

10.2.6 Software, Cloud, Platforms, API, etc.

So far in this project, we have built a device, and it sends messages on the Sigfox network (essentially to the Sigfox servers), now we need to process those messages and do something useful with them.

10.2.6.1 Sigfox Callback

The first thing to do is to have the Sigfox server forward any messages received from your device to some web server/web services that you control. There are many options with the Sigfox system on how to do this but probably the easiest is to build your own RESTful web services and have the Sigfox servers make an HTTP(S) request to your new services with the message data (see the next section). This can be done within the Sigfox backend by using a Callback Mechanism for your device, where you can specify the posted variables or URL parameters as needed from a list of available variables, including the raw message data (Figure 10.10):

Figure 10.10 Example screenshot from Sigfox backend. *Source:* Sigfox.

10.2.6.2 RESTful Web Services

RESTful web services are the modern API. They are ubiquitous on the web, and there are many ways to create them. Here, we show an example of a web service API written in Go (a rapidly growing language used for Web Services currently). The basic structure of a web service (saving to a MongoDB database) in Go looks like this:

```
// Handler for HTTP Post - "/sensordata"
// Register new sensor data
func NewSensorData(w http.ResponseWriter, r *http.
Request) {
    var dataResource Sensor
    // Decode the incoming Task json
    err := json.NewDecoder(r.Body).
Decode(&dataResource)
    if err != nil {
```

```
            common.DisplayAppError(
                w,
                err,
                "Invalid Sensor Data format",
                500,
            )
            return
        }
        sensorData := &dataResource.Data
        context := NewContext()
        defer context.Close()
        c := context.DbCollection("SensorData")
        repo := &db.SensorDataRepository{c}
        // Insert a sensor data document
        repo.Create(sensorData)
        if j, err := json.Marshal(Sensor {Data: *sensor-
Data}); err != nil {
            common.DisplayAppError(
                w,
                err,
                "An unexpected error has occurred",
                500,
            )
            return
        } else {
            w.Header().Set("Content-Type", "applica-
tion/json")
            w.WriteHeader(http.StatusCreated)
            w.Write(j)
        }
}
```

It is worth noting that most of the simple web services you might build for basic data processing of raw data from the Sigfox servers would be of a similar structure.

One thing that would be of particular use for Sigfox message parsing would be the bit unpacking, which will be discussed in further detail in Section 10.2.7.2. Since Sigfox messages are a maximum of 12 bytes, you really need to be squashing as much data as possible into the message, and as such you will probably be bit packing data. The corresponding Go code for unpacking the data, that was bit packed with the earlier Arduino code, looks like this:

```
func bit(n uint64) uint64 {
    return 1<<n
}

func bit_set(y uint64, mask uint64) uint64 {
    return y | mask
}

func bit_clear(y uint64, mask uint64) uint64 {
    return y & ^mask
}

func bit_flip(y uint64, mask uint64) uint64 {
    return y ^ mask
}

func bit_mask(len uint64) uint64 {
    return bit(len) - 1
}

func Bf_mask(start uint64, len uint64) uint64 {
    return bit_mask(len) << start
}

func Bf_prep(x uint64, start uint64, len uint64) uint64 {
    return (x & bit_mask(len)) << start
}

func Bf_get(y uint64, start uint64, len uint64) uint64 {
    return (y>>start) & bit_mask(len)
}

func Bf_set(y uint64, x uint64, start uint64, len uint64)
uint64 {
    return (y & ^Bf_mask(start, len)) | Bf_prep(x,
start, len)
}
```

10.2.7 Microcontroller Coding

Next is writing the basic code to get your breadboarded device to do what you want it to do. Some of this is very standard and included in many of the existing example code for each component, for example, getting the temperature from a DS18B20, covered well in [4], looks like this:

```
#include <DallasTemperature.h>
#include <OneWire.h>
// Data wire is plugged into port 2 on the Arduino
#define ONE_WIRE_BUS 2
// Setup a oneWire instance to communicate with any OneWire
devices (not just Maxim/Dallas temperature ICs)
OneWire oneWire(ONE_WIRE_BUS);
// Pass our oneWire reference to Dallas Temperature.
DallasTemperature temp_sensor(&oneWire);

void setup(){

  Serial.begin(9600);
  temp_sensor.begin();

  Serial.println("DS18B20 Temperature Test\n\n");

  delay(300);//Let system settle

}//end "setup()"

void loop(){
  Serial.print("Requesting temperatures...");
  temp_sensor.requestTemperatures(); // Send the command
to get temperatures

  Serial.print("Temperature is: ");
  float temp_reading = temp_sensor.getTempCBy-
Index(0);
  Serial.println(temp_reading);

  delay(1000);
}// end loop()
```

For low power usage of an Arduino pro mini, there are a number of options in terms of third party libraries. For this project, we will choose the open-source low power library by RocketScream [5] available on GitHub [6]. There is a good article on using this library [7, 8], and the sample usage for this project would be:

```
// **** INCLUDES *****
#include "LowPower.h"
void setup()
{
// No setup is required for this library
}
void loop()
{
// Enter power down state for 8 s with ADC and BOD
module disabled
LowPower.powerDown(SLEEP_8S, ADC_OFF, BOD_OFF);
// Do something here
// Example: Read sensor, data logging, data
transmission.
}
```

10.2.7.1 Sigfox Messages

The Wisol chip chosen for this project can be communicated with using standard AT commands (basic examples are included with the product datasheet). For this project, we need only two functions:

- **Send Message:** A wrapper for the low-level AT commands used for communication with the Wisol chip is included below, this allows for easier command sending such as to test the device and sending messages (here we use the AltSoftSerial library, since the main Serial of the ATMEGA chip is used for communicating debug information with the PC):

```
String send_at_command(String command, int wait_time){
  altSerial.println(command);
  delay(wait_time);
  return recv_from_sigfox();
}
```

```
void test_sigfox_chip(){
  Serial.println("Sigfox Comms Test\n\n");
  altSerial.begin(9600);
  delay(300);//Let system settle

  Serial.println("Check awake with AT Command...");
  chip_response = send_at_command("AT", 50);
  Serial.println("Got reponse from sigfox module: " +
chip_response);
  Serial.println("Sending comms test...");

  chip_response = send_at_command("AT", 50);
  Serial.println("Comms test reponse from sigfox mod-
ule: " + chip_response);

  chip_response = send_at_command("AT$I=10", 50);
  Serial.println("Dev ID reponse from sigfox module: "
+ chip_response);

  chip_response = send_at_command("AT$I=11", 50);
  Serial.println("PAC Code reponse from sigfox module:
" + chip_response);
}
```

- **Enter Low Power (Sleep) Mode:** For this, we opted for the basic sleep mode, though this chip also supports a "deep sleep" option. The rationale behind this choice is that it is not worth moving from ~1.5 to <1 μA as a 1.5 μA Quiescent current drain was more than acceptable for the purposes of this project. The sleep/wake cycle code looks like this:

```
//Sigfox sleep mode enabled via AT$P=1 command
// to wake need to set UART port low (see AX-SIGFOX-
MODS-D.PDF for further details)
void set_sigfox_sleep(bool go_sleep){
  String chip_response;
  if (go_sleep){
    //send go sleep AT command
    chip_response = send_at_command("AT$P=1", 100);
    Serial.println("Set sleep response: " + chip_
response);
  }else{
    //wake up sigfox chip
    altSerial.end();
    pinMode(TX_PIN, OUTPUT);
```

```
        digitalWrite(TX_PIN, LOW);
        delay(100);
        altSerial.begin(9600);
    }
}
```

10.2.7.2 Bit Packing

One thing that would be of particular use for Sigfox message sending is bit packing [9], since Sigfox messages are a maximum of 12 bytes you really need to "stuff" as much data as possible into the message. For example, assume the "temperature" returned by the temperature sensor is going to be a float between −40 and +80 °C, such as 22.46 or −4.67 or something. A float in C++ uses 4 bytes of memory, but you don't want to use up 4 bytes of your 12-byte message sending a number like this if it is not necessary. For most purposes, you only need to know a temperature value to a half degree of accuracy, so if your range of possible temperatures is from −40 to +80 for example, and you only need accuracy to a half degree then you only have 240 possible values you might need to send, so you have squashed them all into 8 bits (1 byte), essentially:

0b00000000 [0] = −40
0b00000001 [1] = −39.5
0b00000010 [2] = −39
...
0b11101111 [239] = 79.5
0b11110000 [240] = 80

For this project, we will choose to use only 7 bits for temperature (−10 to +50 in half degree accuracy), 11 bits for light level (from 0 to 1000 essentially) and a single bit for open/close or device move, and 4 bits for a message sequence number so we can spot any missed messages.

A set of bit packing functions (original code here [10]) is adapted to that we would take all the sensor data as well as the number of bits we want to use for each and pack them into a single 12-byte value:

```
#define BIT(n)                      ( 1UL<<(n) ) //UL =
unsigned long, forces chip to use 32bit int not 16

#define BIT_SET(y, mask)            ( y |=  (mask) )
#define BIT_CLEAR(y, mask)          ( y &= ~(mask) )
#define BIT_FLIP(y, mask)           ( y ^=  (mask) )
/*

          Set bits          Clear bits          Flip bits
y           0x0011            0x0011              0x0011
```

```
mask     0x0101 |           0x0101 &~        0x0101 ^

      ---------        ----------       ---------
result   0x0111            0x0010           0x0110
*/

//! Create a bitmask of length \a len.
#define BIT_MASK(len)                ( BIT(len)-1 )
//! Create a bitfield mask of length \a starting at
bit \a start.
#define BF_MASK(start, len)     ( BIT_MASK(len)<<(start) )

//! Prepare a bitmask for insertion or combining.
#define BF_PREP(x, start, len)   ( ((x)&BIT_MASK(len))
<< (start) )

//! Extract a bitfield of length \a len starting at
bit \a start from \a y.
#define BF_GET(y, start, len)    ( ((y)>>(start)) &
BIT_MASK(len) )

//! Insert a new bitfield value \a x into \a y.
#define BF_SET(y, x, start, len)     \
( y= ((y) &~ BF_MASK(start, len)) | BF_PREP(x, start, len) )

namespace BitPacker {
    static uint32_t get_packed_message_32(unsigned int
values[], unsigned int bits_used[], int num_vals){
        uint32_t retval = 0x0;
        int j = 0;
        for (int i=0;i<num_vals;i++){
            BF_SET(retval, values[i], j, j + bits_
used[i]);
            j += bits_used[i];
        }
        return retval;
    }

    static uint64_t get_packed_message_64(unsigned int
values[], unsigned int bits_used[], int num_vals){
        uint64_t retval = 0x0;
        int j = 0;
        for (int i=0;i<num_vals;i++){
           BF_SET(retval, values[i], j, j + bits_used[i]);
            j += bits_used[i];
        }
        return retval;
    }
}
```

10.2.7.3 IFTTT Integration

In terms of making your device accomplish something beyond data logging, probably the easiest way to integrate it with other devices or ecosystems is to make use of the existing infrastructure for integration and use If This Then That (IFTTT) which is an amalgamation of many different APIs and systems. Once you connect your device to this system, all the existing follow on actions become available. For example, "If [your device sends x] then [send email to y] or [make Alexa say Y] or [Turn on Philips lights in Z room]" or any myriad of other options. There are good references on how best to connect into the IFTTT system, for example, see [11].

10.2.8 From Breadboard to PCB

Once the breadboard, Veroboard, and other prototyping platforms are complete, we have to get onto the PCB design itself. This is where we want to take the working prototype and produce a PCB that can be soldered together into a neat piece of electronics that will achieve the goal of the product. Ultimately something like this is the goal (Figure 10.11):

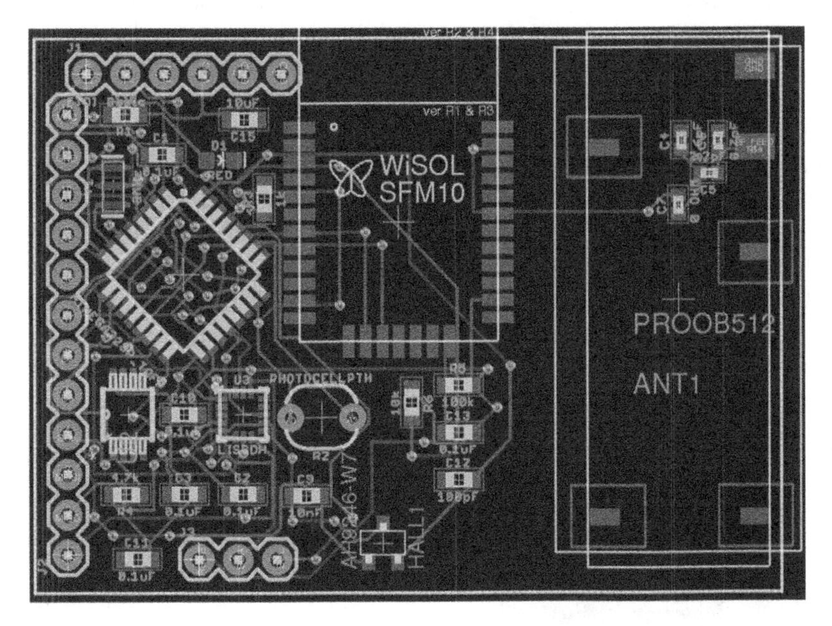

Figure 10.11 Final board layout, fully routed.

The general process is to design the PCB schematic first, and once everything is connected the way it needs to be (i.e.: matching your prototype/breadboard circuit) you will spend time placing all the components on a PCB and routing all the appropriate connections.

There are many different software packages for PCB design, here we use Autodesk Eagle [12], which is an excellent piece of software and free to use for small boards (<80cm). There are lots of component libraries, including third party libraries (for example: all the SparkFun and AdaFruit components).

There are some very useful resources available from SparkFun, including several tutorials for how to get started designing your own PCB based on your breadboard design. Learning how to use this software is too large of a topic for this Chapter, but most of what you need to begin laying out circuits using Eagle can be found in the following set of tutorials, which are highly recommended:

1) Install and setup [13]
2) Creating schematics [14]
3) Board layout and routing [15]

It takes some time to complete all three, but they are well worth it. Some additional tips we would suggest:

- Save often!
- Always run a Design Rule Check (DRC) after every change, no matter how small. Recheck after a "ground pour", or "Ratsnest" command even if the change "should" not have affected the ground connections. It is easy to miss how a single new trace can cut off a ground connection from a ground pour or via, so always run this check after any change before creating Gerber files for PCB printing (see Section 10.2.10 later)
- When routing with very small components (e.g.: FPGA surface mount components), try not to have any holes underneath the component. While this is allowable in manufacturing and should work fine, it becomes an issue when you are hand soldering/surface mounting components for prototype testing in the absence of professional tooling (e.g.: solder reflow ovens, pick & place machinery, etc.). It is hard to be sure when hand applying solder/solder paste that it does not sit under the component and flow into a routing hole underneath (where you cannot see), and it is easy to forgot when routing just how small some of these components are.

  i.e.: Instead of this...

... Do this

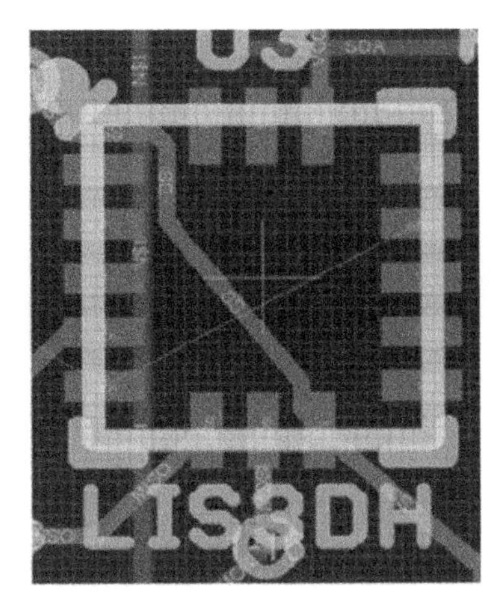

As above, but with larger components, try not to have a via too near any component legs/pads for the same reason (unless the via is also connected to that pad/leg).

Here is the resulting Eagle schematic for this project board (Figure 10.12)

10.2.8.1 Hand Soldering the Surface Mount Components (SMCs)

This can be a daunting question: How to build prototypes that include surface-mounted components? It is clearly much easier to use plated through hole (PTH) components for prototyping (e.g.: breadboarding) but you wouldn't choose PTH components for a final product as SMCs are smaller and neater. So what happens when you design your PCB layout with your ideal SMCs components and you get it printed and you want to put it all together and test it, but you don't have any Surface Mount machinery like a pick and place machine or a solder reflow oven?

Luckily, most SMC components can be hand soldered with a little patience and the right tools (e.g.: a good quality solder iron and a hot air gun).

The YouTube channel EEVBlog covers much of the basics of how to do SMC soldering [16] by hand, and it is certainly possible to hand soldering everything down to 0402 size components (so small you will lose them if you breathe too heavily on them!). See component size comparison chart to understand the different standard package components (Figure 10.13):

It would not be recommended to use 0402 components in your circuit as they are particularly difficult to solder, and in fact 0602 components are way easier to solder while remaining very small and neat. It is recommended that when ordering your PCB printing to order an extra couple of boards in the first batch purely for soldering practice as you will very likely make a mess of your first attempt!

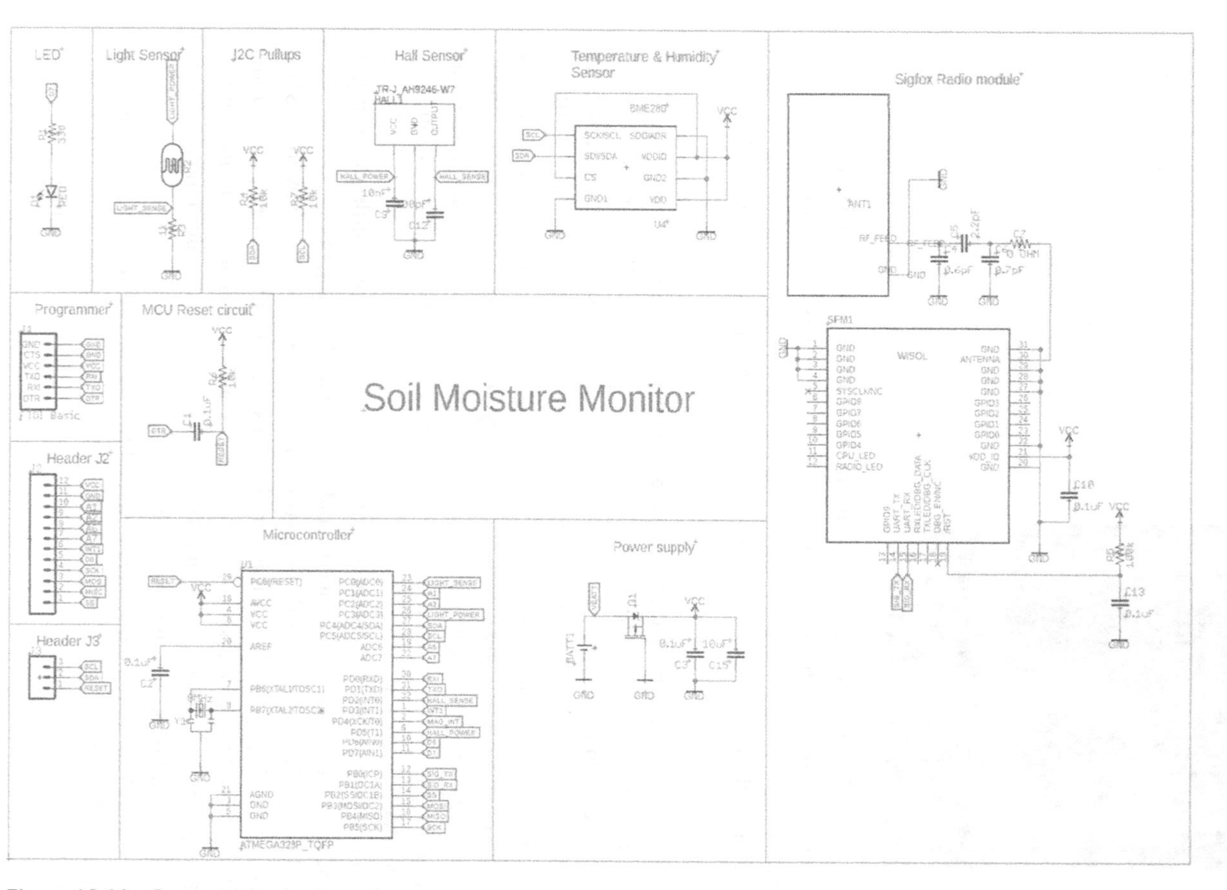

Figure 10.12 Product I final schematic.

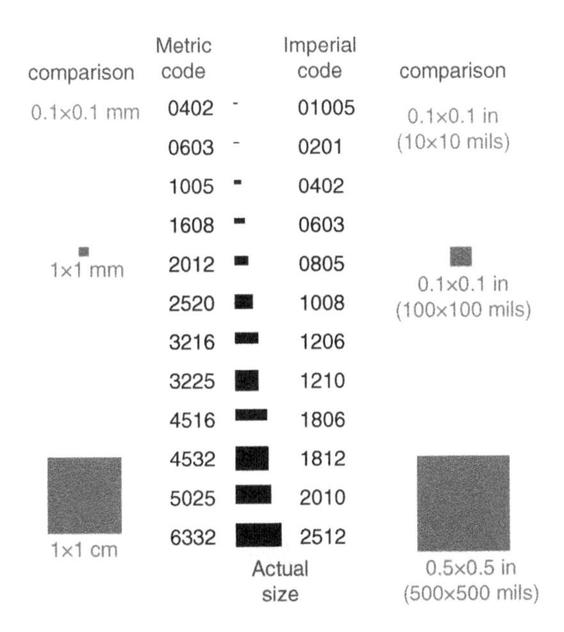

Figure 10.13 Various SMC component sizes.

Finally, in terms of tools needed:

- Soldering iron: a good soldering iron is essential. It is worth paying a bit more for a quality iron as the cheaper ones are just not good enough.
- Hot air soldering gun: This can make soldering some of the smaller VFLGA package ICs like the LIS3DH much easier. It also makes removing components easier when you mess something up.
- Digital multimeter: An essential piece of equipment for any electronics work. It will help with everything from checking for short circuits, voltages, currents, the certain connections are correctly in place, to verifying the correct resistance of components.
- Tweezers: A good quality, fine tip set of tweezers is essential as you will be picking up some very small components.
- Eye loupe/magnifying glass: You will need to be zooming in on your soldering to check for bad solder, solder bridges, blobs, missed pins, etc. A jeweler's loupe, preferably with a built-in light, is especially useful.

10.2.9 Testing and Iteration

Once the Eagle design is completed, and the board is routed, and then sent off to a PCB printer to get printed, you will have to hand solder the resulting board together and start testing. It is likely that the schematic or board layout will need some minor fixes and updates in order for it to work exactly as intended. It is important to label your board with a version number for each printing, as they will often look very similar when printed as some iterations can be only minor changes, or even simply routing changes that are hard to see.

Here is a soil moisture monitor assembled PCB and test setup (Figure 10.14):

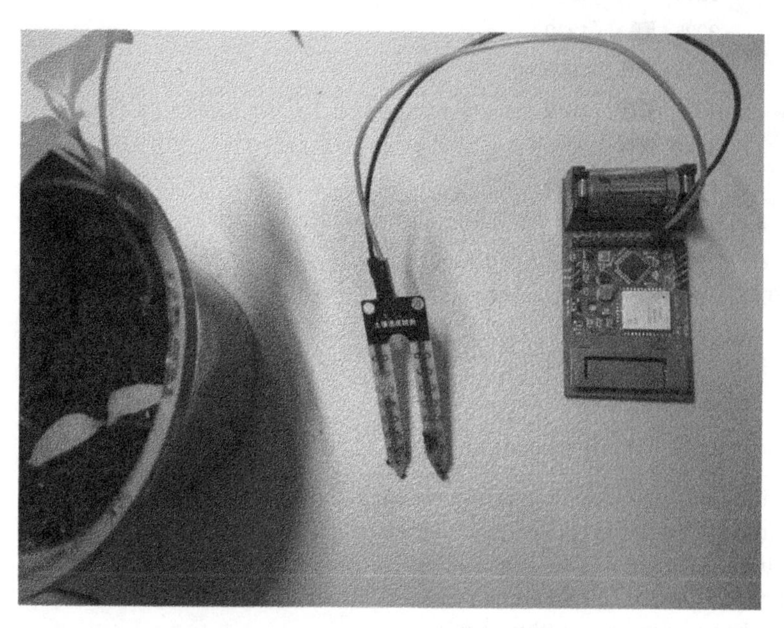

Figure 10.14 Testing soil moisture level and sensor processing.

During testing, it was found that the simple temperature sensor (the DS18B20) did not give reliable enough readings, and it was felt that a more advanced module that also included other weather data (like accurate humidity and atmospheric pressure) would be a better long-term solution. So, the schematic was updated and the DS18B20 was replaced with a Bosch BME280. This emphasizes both the importance of testing devices in the field and not to be afraid of making changes to a design to improve wide use cases for the device.

The interface for extracting useful information from the Bosch sensor is a little more complicated, but luckily, there is a good library from Adafruit that does all the hard work for the user, available here [17]. For example, here is how to get simple temperature readings from a Bosch BME280 (code taken from the Adafruit Github linked above):

```
/*****************************************************
This is a library for the BME280 humidity, temperature
  & pressure sensor
Designed specifically to work with the Adafruit BME280
  Breakout
----> http://www.adafruit.com/products/2650
These sensors use I2C or SPI to communicate, 2 or 4
  pins are required
to interface. The device's I2C address is either 0x76
  or 0x77.
Adafruit invests time and resources providing this
  open source code,
please support Adafruit and open-source hardware by
  purchasing products from Adafruit!
Written by Limor Fried & Kevin Townsend for Adafruit
  Industries.
BSD license, all text above must be included in any
  redistribution
See the LICENSE file for details.
*****************************************************
#include <Wire.h>
#include <SPI.h>
#include <Adafruit_Sensor.h>
#include <Adafruit_BME280.h>

#define BME_SCK 13
#define BME_MISO 12
#define BME_MOSI 11
#define BME_CS 10

#define SEALEVELPRESSURE_HPA (1013.25)

Adafruit_BME280 bme; // I2C

unsigned long delayTime;

void setup() {
    Serial.begin(9600);
    while(!Serial);    // time to get serial running
    Serial.println(F("BME280 test"));
```

```
    unsigned status;
    // default settings
    status = bme.begin();
    if (!status) {
        Serial.println("Could not find a valid BME280
sensor");
        while (1) delay(10);
    }
    Serial.println("-- Default Test --");
    delayTime = 1000;
}

void loop() {
    printValues();
    delay(delayTime);
}

void printValues() {
    Serial.print("Temperature = ");
    Serial.print(bme.readTemperature());
    Serial.println(" *C");

    Serial.print("Pressure = ");
    Serial.print(bme.readPressure() / 100.0F);
    Serial.println(" hPa");

    Serial.print("Approx. Altitude = ");
    Serial.print(bme.readAltitude(SEALEVELPRESSURE_
HPA));
    Serial.println(" m");

    Serial.print("Humidity = ");
    Serial.print(bme.readHumidity());
    Serial.println(" %");

    Serial.println();
}
```

And the code used during the testing of the soil moisture sensor specifically (taken from the SparkFun soil sensor guide [18]):

```
/*  Soil Mositure Basic Example
    This sketch was written by SparkFun Electronics
    Joel Bartlett
    August 31, 2015
Basic sketch to print out soil moisture values to the
Serial Monitor
    Released under the MIT License(http://opensource.
org/licenses/MIT)
*/

int val = 0; //value for storing moisture value
int soilPin = A0;//Declare a variable for the soil
moisture sensor
int soilPower = 7;//Variable for Soil moisture Power

void setup()
{
  Serial.begin(9600);   // open serial over USB

  pinMode(soilPower, OUTPUT);//Set D7 as an OUTPUT
  digitalWrite(soilPower, LOW);//Set to LOW so no power
is flowing through it
}

void loop()
{
  Serial.print("Soil Moisture = ");
  //get soil moisture value from the function below
and print it
  Serial.println(readSoil());

  delay(1000);//take a reading every second
}
//This is a function used to get the soil moisture content
int readSoil()
{

    digitalWrite(soilPower, HIGH);//turn D7 "On"
    delay(10);//wait 10 milliseconds
    val = analogRead(soilPin);//Read the SIG value
form sensor
    digitalWrite(soilPower, LOW);//turn D7 "Off"
    return val;//send current moisture value
}
```

10.2.10 PCB to Finished Product

At various stages, from breadboarding to bulk manufacturing, you will need to make use of a variety of resources:

- **Hardware components:** To breadboard your circuit, you will need components such as resistors, capacitors, sensors, integrated circuits, etc. You can find some of these in mainstream sites like Amazon, but it is recommended to use some of the hardware-specific sites such as DigiKey, Mouser, or Farnell.
- **Gerber files:** In order for a PCB manufacturer to print your PCB, you will need to send them the files in the format they need for manufacturing, which is typically a set of "Gerber" files. The Gerber files are an open ASCII format for printed circuit boards which include the necessary information like Pad sizes, drill sizes, top and bottom copper traces, and pours, etc., that will allow a PCB manufacturer to produce this board. In Eagle, the software used in this project, there is support for automatically creating the Gerber format from your board by choosing the computer-aided manufacturing (CAM) processor option from the menu bar. Then you should either use the default settings, or better yet, ask your manufacturer for their CAM settings file and use that instead, as then you will know for sure your board can be printed without any issues by your chosen manufacturer.
- **PCB printing:** Once you have designed your PCB and created the Gerber files, you will need someone to print this for you. Multi-CB in Europe are particularly good, efficient as well as very competitively priced, and Advanced Circuitry International and Onboard Circuits if you are in the USA. Depending on where you are based, there may be other PCB manufacturers available to you, there is a section in [19] under "Picking a PCB Manufacturer" that might be worth a look.
- **Compliance:** It may be necessary to put your product forward for various tests to be compliant with laws around resale of electronic devices in the countries you plan to market your device. Some likely areas of compliance to be aware of:
  - **FCC (US):** Ensuring your device meets the necessary RF standards where it uses radio frequencies in certain ranges.
  - **CE mark (Europe):** A certification mark that indicates your device conforms to EU health and safety laws. There are many firms that can help you gain this certification.
  - **Sigfox certification:** If using the Sigfox network, as the device in this project does, you will need to gain Sigfox certification in order to sell your device as a "Product" on their website. This process also ensures that your device has sufficient power and reliability to send messages that will be received properly by the Sigfox base stations.

- **Enclosure:** A finished PCB is the most important part of the process, but in real terms, to use a product like this in the field, it will need to be protected from the elements within some kind of enclosure. It is possible, through a number of small project websites, to engage a 3D modeler to create a set of CAD files for a design to fit your product and to your specifications. For example, below is a 3D model of the design that will fit the product built in this section, which can be 3D printed online, or in bulk (Figure 10.15):

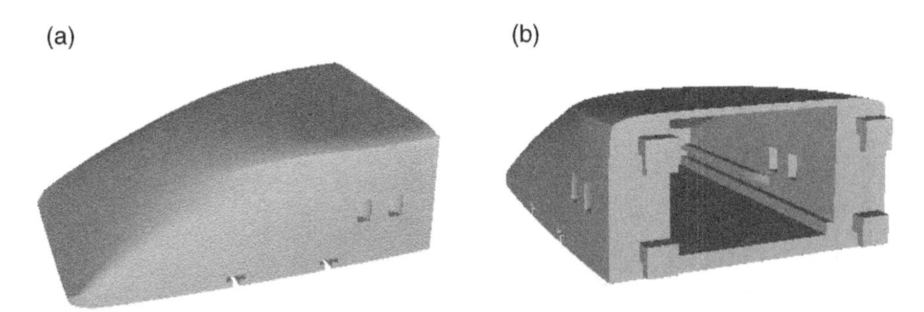

Figure 10.15 Enclosure, side view (a), rear view (b).

However, it will often be less expensive to find an existing, workable enclosure of the right size, and tailor your PCB to fit it instead (e.g. by adding mounting holes to the PCB at the correct points). We will leave it up to the reader to decide on this as it depends largely on how the end product should look and the acceptable final cost.

- **PCB manufacturing:** Once your PCB has been printed, everything is soldered together, tested and all is working perfectly, the next consideration might be where to have a large number of boards manufactured. PCB Cart in China is good and reasonably priced. The price can include assembly and the programming of the Atmega chip with a default program.
- **BOM:** You will also need to produce a Bill of Materials, or BOM, for your product that details exactly how many of each component, and their designation on your schematic, are needed to produce the final product. This also gives you a good idea of the per unit cost of the device, on top of the per unit manufacturing cost. For example, the BOM for the product built in this section is below, giving materials input total cost per device (per 100 devices manufactured) of €13.140.

Item	Schematic ref	Device	Quantity	Manufacturer P/N	Indicative cost (per 100)	Total cost
1	C1 C2 C3 C8 C10 C11 C13	0.1UF-0603-25V-5%	7	CL10F104ZA8NNNC	€0.0065	€0.046
2	C4	0.6PF-0402-50V-0.1PF	1	GRM1555C1HR60WA01D	€0.0528	€0.053
3	C5	2.2PF-0402-50V0.1PF	1	GRM1555C1H2R2WA01D	€0.0308	€0.031
4	C6	0.7PF-0402-50V-0.1PF	1	GRM1555C1HR70WA01D	€0.0528	€0.053
5	C7	RESISTOR0402	1	RC0402JR-070RL	€0.0032	€0.003
6	C9	10NF-0603-50V-10%	1	GRM188R71H103KA01D	€0.0570	€0.057
7	C12	100PF-0603-50V-5%	1	GRM1885C1H101JA01D	€0.0581	€0.058
8	C15	10UF-0603-6.3V-20%	1	GRM188C81C106MA73D	€0.1353	€0.135
9	D1	LED-RED0603	1	LTST-C191KRKT	€0.0943	€0.094
10	HALL1	AH9246-W7	1	AH9246-W-7	€0.3006	€0.301
11	J1	6_PIN_SERIAL_TARGET	1	GRPB061VWVN-RC	€0.3170	€0.317
12	J2	CONN_12	1	PEC12SAAN	€0.4693	€0.469
13	J3	CONN_03	1	GRPB031VWVN-RC	€0.1656	€0.166
14	R1	330OHM-0603-1/10W-1%	1	RC0603FR-07330RL	€0.0051	€0.005
15	R2	PHOTOCELLPTH	1	PDV-P8103	€0.3904	€0.390

16	R3	1KOHM-0603-1/10W-1%	1	RC0603FR-071KL	€0.0051	€0.005
17	R5	100KOHM-0603-1/10W-1%	1	RC0603FR-07100KL	€0.0051	€0.005
18	R4 R6 R7	10KOHM-0603-1/10W-1%	3	RC0603FR-0710KL	€0.0051	€0.015
19	SFM1	WISOL Sigfox Chip	1	WSSFM10R1	€2.8400	€2.840
20	U1	ATMEGA328P_TQFP	1	ATMEGA328P-AU	€1.4884	€1.488
21	Y1	RESONATOR-8MHZSMD_3.2X1.3	1	CSTCE8M00G55-R0	€0.2734	€0.273
22	ANT1	PRO-OB-471	1	PRO-OB-471	€1.0446	€1.045
23	BATT1	CR2-BATTERY-HOLDERPTH	1	BH-CR2-PC	€1.3900	€1.390
24	Q1	MOSFET P-CH 20V 5.3A SOT-23	1	SI2323DDS-T1-GE3	€0.3600	€0.360
25	U4	BME280	1	BME280	€3.5400	€3.540
					Total:	***€13.140***

10.3 Product II (Wearable): Fall Detection Device

The idea of this example project is to design a wearable product that can reliably detect an elderly person's accidental fall, and then, if a fall is detected, the device would promptly contact emergency for help. A smartphone application should enhance the product's capabilities by offering a cloud connectivity, which would make use of a cloud solution to log the events. It would also potentially extend the product use case based on future requirements, for instance, updating the device location information.

10.3.1 Product Requirements and Design Considerations

As mentioned in Chapter 6, the requirements are documented and serve as an agreement between the client and the product engineering team. The document can be used towards the product delivery time as a checklist for product completeness upon delivery.

In this sample wearable project, a subset of some important system requirements and design considerations is picked as shown below:

- **Accuracy**: The system should accurately detect sudden falls
- **Usability**: The system should be relatively easy to configure and use
- **Size**: The system should be small, preferably less than 2″ × 2″ × 0.5″ (L × W × H)
- **Weight**: The system should be light, preferably lighter than 50 g (1.77 oz.)
- **Power**: The system should maintain operation for at least eight hours per charge

10.3.2 Design Block Diagram

The following diagram shows the end-to-end building blocks of the wearable solution, catered to the use case described in this section (Figure 10.16).

- **Wearable device**
  - **Microprocessing unit (MPU):** The MPU periodically runs an algorithm to process accelerometer and gyroscope data to detect fall events. It also manages the communication with the user's smartphone over BLE.
  - **Fall detection sensors:** e.g., accelerometer and gyroscope sensors to detect sudden changes in acceleration and orientation.
  - **Bluetooth Low Energy (BLE) module:** Interfaces with smartphone to send fall detection alarm signals to the intended party.
- **Smartphone:** Receives fall detection alarm signals from the wearable device, contacts emergency, and provides Internet connectivity to the cloud to log events.
- **Cloud:** Event logging and potential reactive capability (i.e.: event-based or criteria-based action triggers).
- **User:** A person wearing the fall-detection device

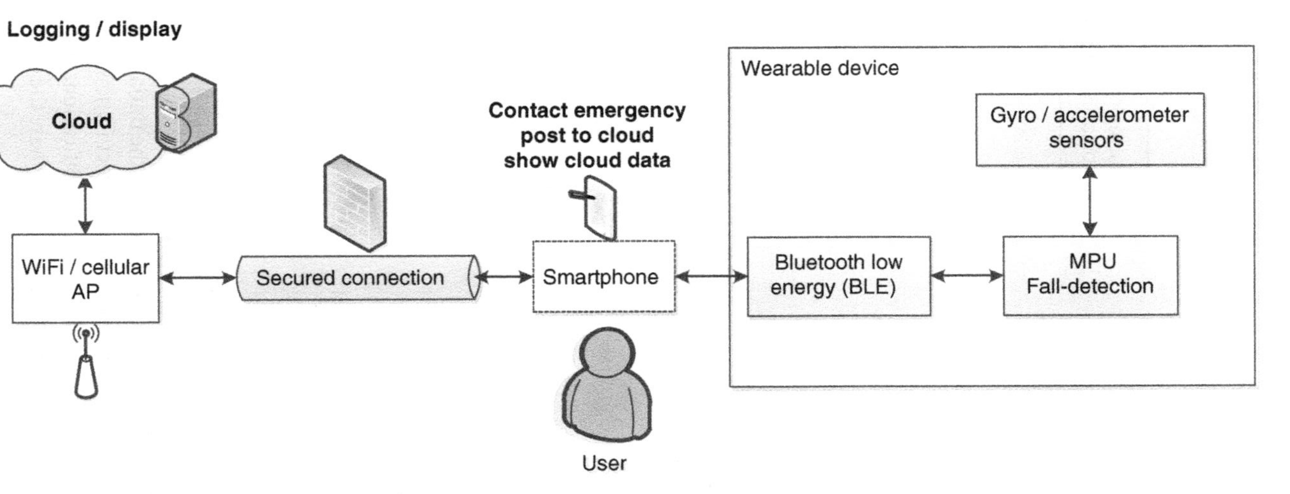

Figure 10.16 Fall-detection system block diagram.

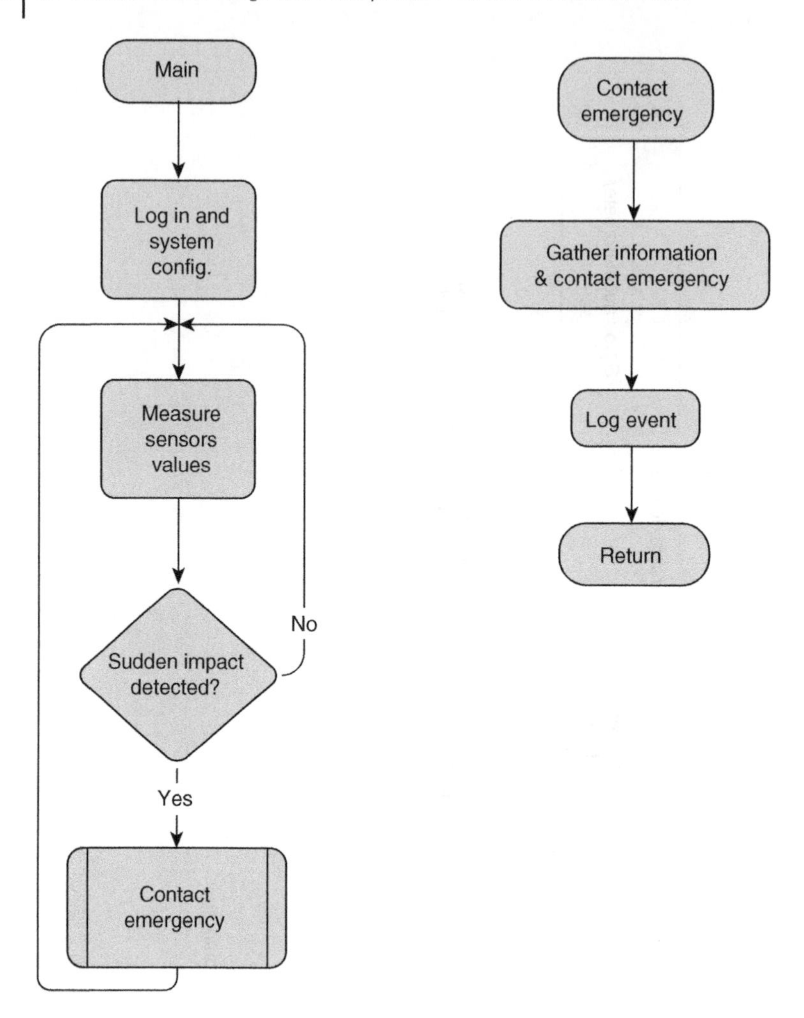

Figure 10.17 Program execution flowchart.

10.3.3 Flowchart

A flowchart is a good way to illustrate the execution steps the system software will take in a two-dimensional format, especially in visualizing conditional execution and function calls. Flowcharts are equally useful at the initial system design stages as well as the final documentation stage of a project (when the system is operational) in order to aid in its use, maintenance, modification, and expansion.

Figure 10.17 shows the software execution steps where the execution starts with:

1) System powers up.
2) User enters their credentials to login to the system.

3) System configuration starts which entails setting up the input/output interfaces, the communication links, and the emergency contact info.
4) Measurement routine starts, and the sensors values are periodically evaluated to detect a user falling event.
5) Once a fall-detection event is asserted, the device will trigger the contact emergency subprocess.
6) As part of the contact emergency sub-process, a text message will be sent to the preconfigured emergency contact and the event will be logged to the cloud.
7) This system will be running forever until it is reset or powered off, this is a typical process for embedded systems.

10.3.4 Unified Modeling Language (UML)

Unified modeling language is a standardized general-purpose modeling language that is also used for developing software in a visual way. In this example, the UML was used to visually illustrate the product use case and how the components relate to each other. The use case UML diagram in Figure 10.18 is translated from the system diagram and flowchart to describe the user and system components interactions.

10.3.5 Hardware Selection

As seen from the system requirements section, the device of choice should be on the lower side of the power consumption, has a small form factor, light in weight,

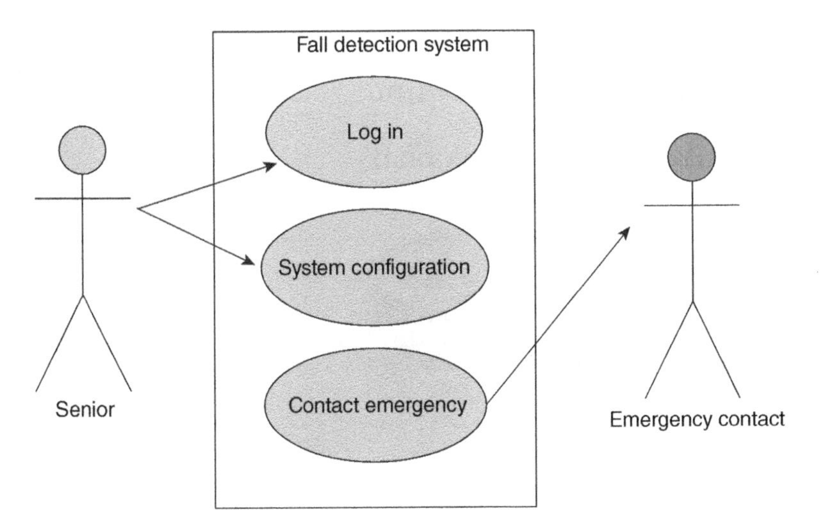

Figure 10.18 Use case UML diagram.

accurate in its reading of sensor measurements and ultimately detecting a fall event, and relatively easy to setup and use.

There are numerous off-the-shelf development boards available, and each has its own strength points and uses. For this example, the Mikroelektronika-NXP Hexiwear is used. This modular development board can be a complete development solution. It uses the power efficient 120 MHz K64F 32-bit ARM cortex M4 processor and has an elegant smartwatch design. It also has six accurate sensors including the accelerometer and gyroscope sensors which are used in this project. It also includes a chip for BLE connectivity to communicate with the smartphone and has the flexibility to add more modules if a docking station is used. Figure 10.19 shows the Hexiwear wearable development board.

The Hexiwear accelerometer and gyroscope sensors are used to get the raw measurements for the user's movements. The firmware on K64f MPU runs an algorithm to detect a fall and trigger an alarm which will be communicated to the smartphone app that will in turn trigger a preconfigured SMS message.

It is important to understand that picking the right development board that has most, if not all, of the components needed for a prototype can save a tremendous amount of time during the hardware and software implementation phases later, as will be seen in the following sections.

The Hexiwear development board features more than traditional MCU development platforms. It is ideal for connected applications since it includes the following components:

- Kinetis K64F MCU
- Bluetooth Low Energy (BLE) SoC (Kinetis KW40z)
- 6-axis Accelerometer and Magnetometer combo (FXOS8700CQ)
- 3-axis Gyroscope (FXAS21002CQ)
- Pressure sensor accurate up to Altitude sensing (MPL3115A2)
- Temperature and humidity combo (HTU21D)
- Ambient light sensor (TSL2561)
- Optical Heart rate sensor (Maxim MAX30101)

Figure 10.19 Hexiwear wearable development board. *Source:* Photo courtesy of MikroElektronika.

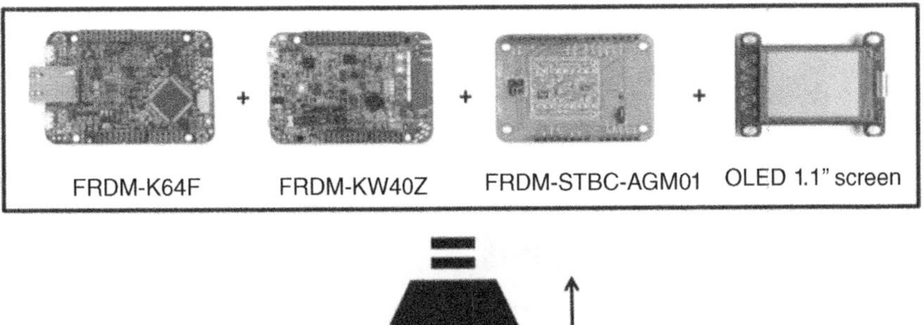

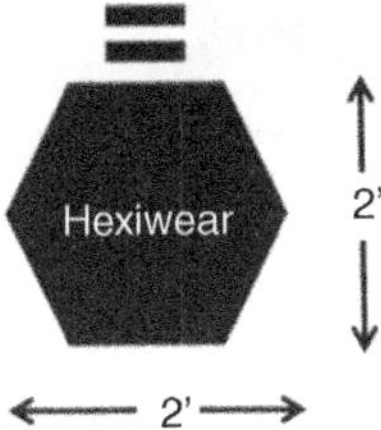

Figure 10.20 Development boards constituting the Hexiwear platform. *Source:* Photo courtesy of MikroElektronika.

- 1.1″ OLED color display
- 190 mAh 2C Li-Po battery with a charger (MC34671)

The platform combines the functionality of (FRDM-K64F, FRDM-KW40Z, FRDM-STBC-AGM01, and OLED 1.1″ Screen) development boards into a single compact module. Figure 10.20 shows the different development boards constituting the Hexiwear platform.

Figure 10.21 shows a block diagram of the components inside the Hexiwear development platform and the optional components if a docking station is used. The docking station is used to program the Hexiwear and debug the running firmware.

Figure 10.22 shows a list of the MPUs capabilities and their peripheral connections.

Figure 10.23 lists the sensors and their specifications and communication interfaces.

10.3.6 Hardware Implementation and Connectivity

As seen from the previous section, the Hexiwear has all the MPUs, sensors, and wireless connectivity modules needed for this example project on the same development board. This choice has the substantial advantage of eliminating the need to add any external components along with the breadboard and wiring which in turn might add development complexity to the hardware implementation (and software complexity as will be seen in the following section). Having the hardware

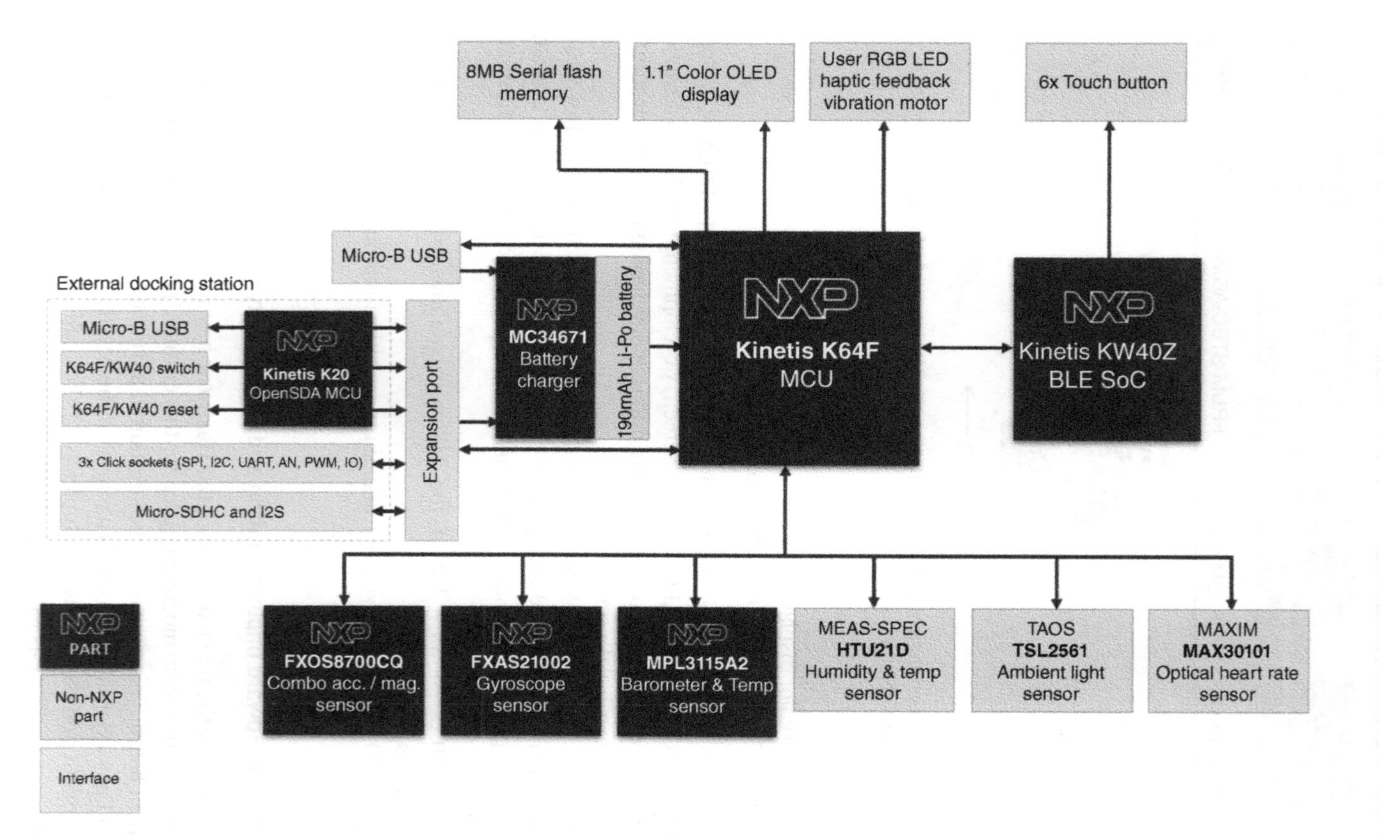

Figure 10.21 Hexiwear default and optional components. *Source:* Courtesy of ARM.

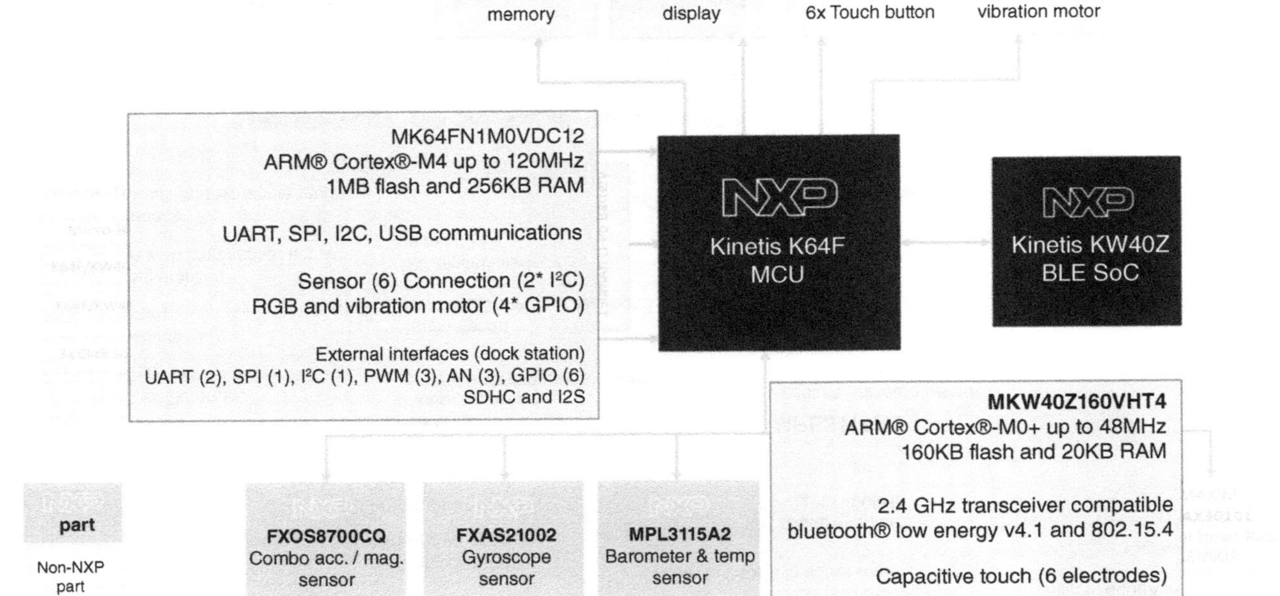

Figure 10.22 Hexiwear MPUs capabilities and their peripheral connections. *Source:* Courtesy of ARM.

HTU21D
Fully-calibrated humidity sensor
+/−3%RH tolerance @55%RH
Fully-calibrated temperature tensor
±0.3°C accuracy from −40 to + 125°C
Consumption
down to 450μA in active mode
12C digital interface
up to 400kHz dual-mode
TSL2561
Light to digital converter
0.1 to 40,000 Lux dynamic range
inc. both infrared and full spectrum diodes
Consumption
down to 240μA in active mode
12C digital interface
up to 400kHz
MAX30101
Pulse oximeter and hear rate sensor
high sensitivity with 16-bit ADC
Consumption
down to 600μA in active mode
12C digital interface
up to 400kHz
FXOS8700CQ
3-axis linear accelerometer
±2 g/±4 g/±8 g dynamic range
3-axis magnetometer
±1200 μT range
Low-power consumption
down to 80 μA with both sensor active
I2C digital interface
up to 400 Hz dual, 800 Hz single-mode
FXAS21002CQ
3-axis gyroscope
±250/500/1000/2000°/s dynamic range
Consumption
down to 2.7 mA in active mode
12C digital interface
up to 800 Hz
MPL3115A2
Absolute pressure sensor
calibrated 50–110 kPa range
altitude accuracy down to 0.1 m
Consumption
down to 8.5 μA (capt.) max 2 mA (with conv.)
Autonomous data-logging
32-sample FIFO up to 12 days
12C digital interface
up to 400 Hz
NXP
FXOS8700CQ
Combo acc. / mag. sensor
NXP
FXAS2100CQ
Gyroscope sensor
NXP
MPL3115A2
Barometer & temp sensor
MEAS-SPEC
HTU21D
Humidity & temp sensor
TAOS
TSL2561
Ambient light sensor
MAXIM
MAX30101
Optical heart rate sensor

Figure 10.23 Hexiwear sensors specification and communication interfaces. *Source:* Courtesy of ARM.

implementation ready out of the box gives an advantage when it comes to the time-to-prototype and hence to market.

10.3.6.1 Hardware Modules and Interfaces Overview

As shown in Figure 10.24, the K64F MCU ARM® Cortex®-M4F interfaces with both the 3-axis accelerometer sensor (FXOS8700) and the 3-axis Gyroscope sensor (FXAS21002) using the interintegrated circuit (I^2C) interface, denoted by I2C1.

The K64F interfaces with the K40 over the Universal Asynchronous Receiver Transmitter (UART) interface, which in turn relays to the outside world using the on-chip BLE module. This is going to be used to communicate with the smartphone app which acts as a gateway to the outside world. As mentioned in Chapter 5, BLE is a more power efficient version of Bluetooth, it is also more energy efficient than ZigBee and classic Wi-Fi. It provides a reliable short range of up to 400 m, and a data rate of 2 Mbit/s. This makes BLE a popular protocol of choice for IoT wireless communications, especially for PANs. This Hexiwear connection with the smartphone is an example of a PAN. The smartphone uses the cellular service to send the emergency text message and uses the Wi-Fi or cellular service to post the data to the cloud.

Figures 10.25–10.28 show the Hexiwear fall-detection wearable device, Hexiwear docking station, Hexiwear beside an apple watch for the purpose of size comparison, and a Hexiwear worn by a user, respectively:

10.3.7 Software Implementation

Development boards in general are designed with the time-to-prototype in mind. That means they come with all the open-source libraries necessary to start the development out of the box. Since the board of choice is manufactured by NXP, and many of the low-level drivers for the hardware components are available as part of the Kinetis Software Development Kit (KSDK), the Kinetis Design Studio (KDS) IDE is going to be used to build and run the firmware source code.

A link to the Hexiwear user manual, which includes the steps to install the IDE and SDK, and the steps to import projects are provided in [20] in the "References" section.

It is usually easier to start with a working example program than to start from scratch. And it is part of the software/firmware developer's responsibility to find a suitable example that works, and then modify/improve it to accommodate the specific use case of their project.

In the fall-detection sample project, a working example is used and modified to have the K64F Cortex M4 processor read the two sensors values (accelerometer and gyroscope), run the fall-detection algorithm, and communicate alarms over BLE. By design, the stock firmware, provided with the Hexiwear, comes with an

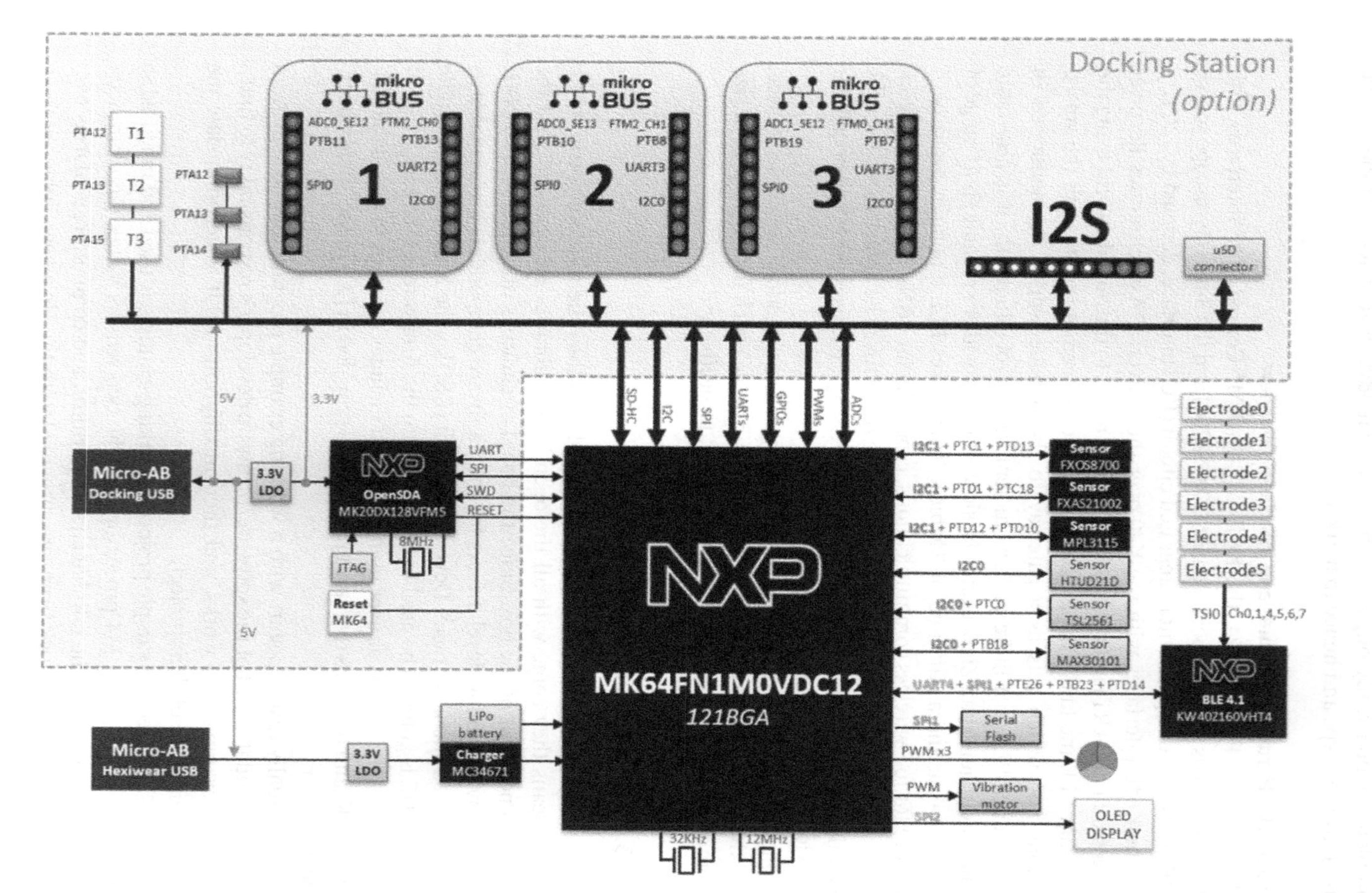

Figure 10.24 Hexiwear K64F interfaces with internal docking station sensors. *Source:* Courtesy of ARM.

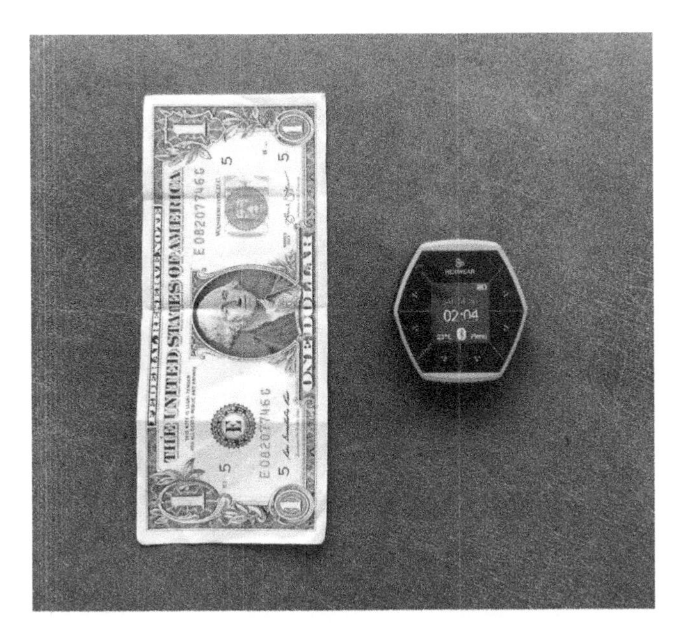

Figure 10.25 (Hexiwear) fall-detection wearable device.

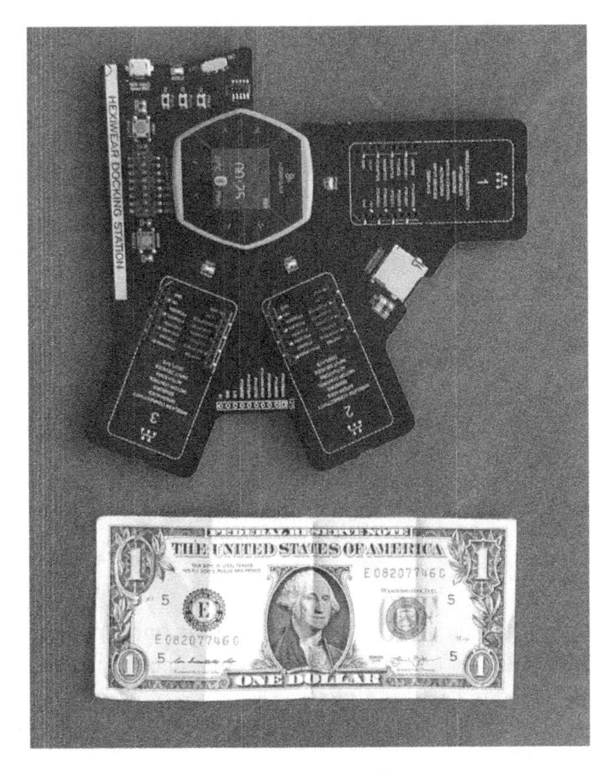

Figure 10.26 Hexiwear docking station.

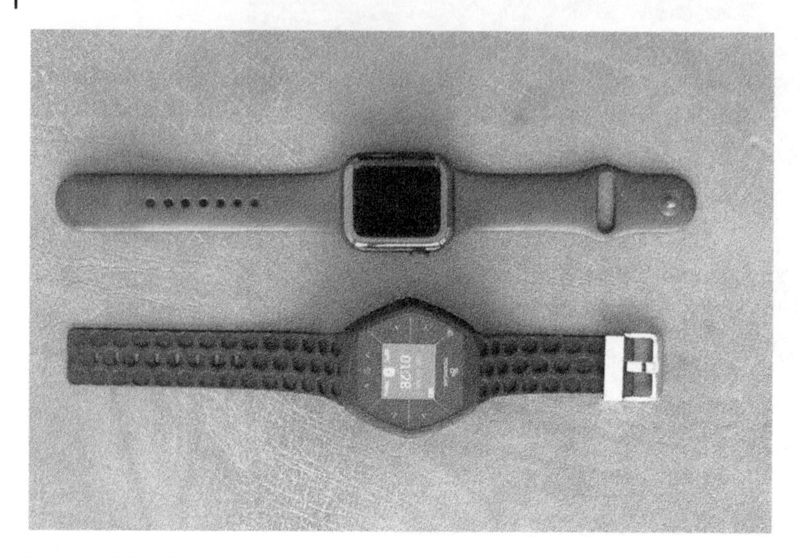

Figure 10.27 Hexiwear beside an apple watch.

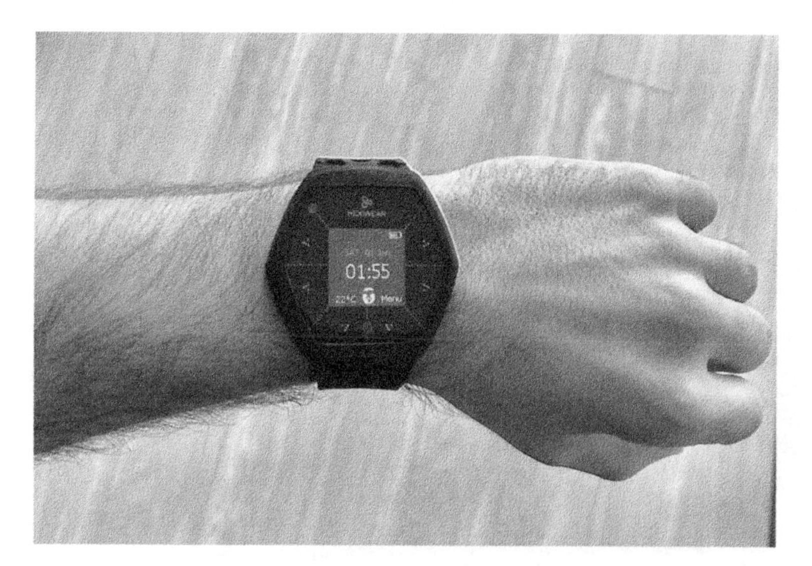

Figure 10.28 Hexiwear worn.

example of a working prototype that fetches readings from the six onboard sensors and output to the OLED and BLE modules. A link to the stock firmware binaries and source code is provided in [21].

Once the IDE software and the SDK are installed and the project is imported, it should look like the snapshot in Figure 10.29:

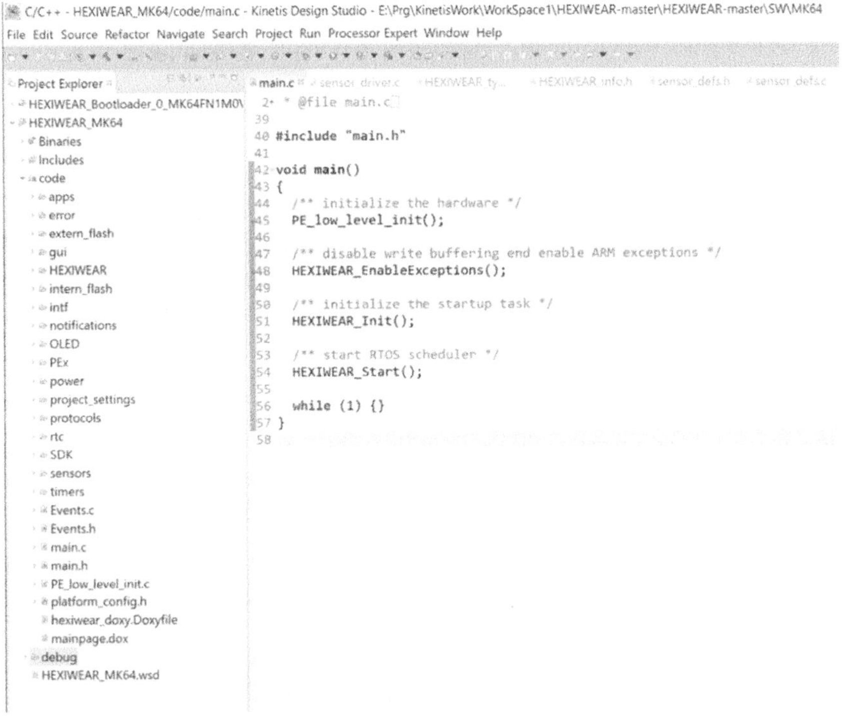

Figure 10.29 KDS stock firmware project import. *Source:* [22].

The control flow starts at the main function, 'void main()', in main.c, by initializing the low-level Hardware and the OSA real-time operating system (RTOS). Second comes the activation of the Cortex M3 exceptions (events and interrupts). The control then goes to the third function, 'HEXIWEAR_Init()', which configures multiple interfaces including the output GPIO pins (OLED, flash, power, VIBRO, KW40, other LEDs), input GPIO pins (battery and tap). The 'HEXIWEAR_Init()' function also initializes the task to continuously read sensors data, initializes the accelerometer sensor, turns on the sensor tag and haptic feedback, initializes structures necessary for the RTOS run, and finally it turns on the system clock.

The last function, 'HEXIWEAR_Start()', starts the RTOS scheduler and the system runs forever until powered off (a typical embedded system operation).

Our objective here is the following:

1) Find the point in code where accelerometer and gyroscope data are being read from hardware.
2) Come up with a relatively short real-time algorithm that detects a fall based on the values read.
3) Once a fall is detected, an alarm signal is sent to the smartphone via BLE.

For point (1), a good probing point for the sensors readings is just before sending the formatted data to the BLE. The location is in file 'sernsor_drive.c', at function 'sensor_PushData()'.

10.3.7.1 Fall Detection Algorithm

It is a good practice to explore the studies that have been published regarding these kinds of algorithms. In order to choose an adequate fall-detection algorithm, the firmware developer needs to verify that the algorithm of choice (or a modified version of it) is implementable on the MPU of the development board. In this example, the selection was to use an algorithm that has already been tested and verified to work as part of a study published in [22]. The study reports that the algorithm accuracy is as high as 100% in the best-case scenario (walking), and 86.67% in the worst-case scenario (going downstairs). Taking that into consideration, this would be a reasonably accurate algorithm to implement.

Equations (10.1)–(10.5) show the algorithm mathematical calculations:

Based on the value generated by the accelerometer, axis made on the magnitude of these axes were denoted as:

$$AT_t = \sqrt{aX_t^2 + aY_t^2 aZ_t^2} \tag{10.1}$$

Meanwhile, the gyroscope applies the same formula as:

$$GT_t = \sqrt{gX_t^2 + gY_t^2 gZ_t^2} \tag{10.2}$$

After discovering the magnitude of the sensor, the next step is to find the maximum and minimum value of the sensor. Below is the formula to find the maximum and minimum values:

$$\text{MAX}\left[AT_t\ AT_{t-n}\right] \text{and MIN}\left[AT_t\ AT_{t-n}\right] \tag{10.3}$$

$$\text{MAX}\left[GT_t\ GT_{t-n}\right] \text{and MIN}\left[GT_t\ GT_{t-n}\right] \tag{10.4}$$

Once the maximum and minimum values are obtained, the following is the formula to find the value sought:

$$\text{Angle} = \arccos\left(\frac{\sqrt{aX_t^2 + aY_t^2}}{aZ_t^2}\right) \tag{10.5}$$

g is the constant of gravity that is $9.8\,\text{m/s}^2$

Figure 10.30 shows the flowchart of the adopted algorithm:

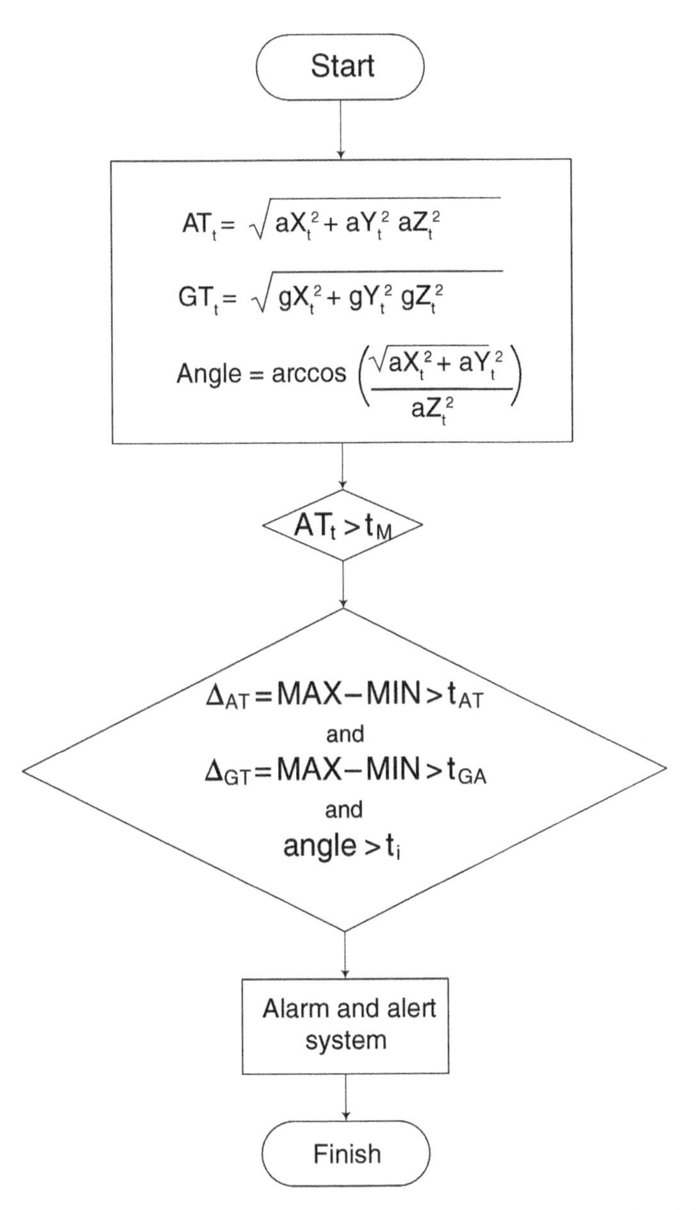

Figure 10.30 Fall-detection algorithm flowchart. *Source:* Reproduced from [22].

The following code snippets show the algorithm implementation:

```
/**
 * Detect a fall and alarm smartphone
 * depending on the sensors's values
 */
static inline void detect_Fall(hostInterface_packet_t
sensorPacket){
  if ((packetType_accel == sensorPacket.type) ||
    (packetType_gyro == sensorPacket.type)){
  static int64_t accel_x_pwr2 = 0;
  static int64_t accel_y_pwr2 = 0;
  static int64_t accel_z_pwr2 = 0;
  static int64_t max_a_t = INT64_MIN;
  static int64_t min_a_t = INT64_MAX; // max value
  static int64_t a_t_i = 0;  //ATi
  static uint64_t delta_a_t = 0;
  static bool acc_exceeded_thrs = false;
  static uint32_t accel_sample_count = 0;
  static int16_t angle = 0;
  static int64_t gyro_x_pwr2 = 0;
  static int64_t gyro_y_pwr2 = 0;
  static int64_t gyro_z_pwr2 = 0;
  static int64_t max_g_t = INT64_MIN;
  static int64_t min_g_t = INT64_MAX; // max value
  static int64_t g_t_i = 0;  //ATi
  static uint64_t delta_g_t = 0;
  static uint32_t gyro_sample_count = 0;
  static uint32_t gyr_exceeded_thr = 0;
  if(packetType_accel == sensorPacket.type){
    accel_x_pwr2 = motionData.accData[0] * motionData.
accData[0];
      accel_y_pwr2 = motionData.accData[1] * motion-
Data.accData[1];
      accel_z_pwr2 = motionData.accData[2] * motion-
Data.accData[2];
      a_t_i = sqrt(accel_x_pwr2 + accel_y_pwr2 +
accel_z_pwr2); // ATi
      if (accel_sample_count < 10000) { // Calibration
      max_a_t = ( a_t_i > max_a_t )? a_t_i : max_a_t;
```

```
      min_a_t = ( a_t_i < min_a_t )? a_t_i : min_a_t;
      accel_sample_count++;
        }else{ // check for threshold
          if (a_t_i > 9){ // TODO: Get a more accurate
value for tm from experiment
            delta_a_t = max_a_t = min_a_t;
            if (delta_a_t > 4.2) // TODO: Get a more
accurate value for tAT from experiment
            acc_exceeded_thrs = true;
            }else{
              acc_exceeded_thrs = false;
              }
        }
  }
  if (packetType_gyro == sensorPacket.type) {
      gyro_x_pwr2 = motionData.gyroData[0] * motion-
Data.gyroData[0];
      gyro_y_pwr2 = motionData.gyroData[1] * motion-
Data.gyroData[1];
      gyro_z_pwr2 = motionData.gyroData[2] * motion-
Data.gyroData[2];
      g_t_i = sqrt(gyro_x_pwr2 + gyro_y_pwr2 + gyro_z_
pwr2); // GTi
      if (gyro_sample_count < 10000) { // Calibration
      max_g_t = ( g_t_i > max_g_t )? g_t_i : max_g_t;
      min_g_t = ( g_t_i < min_g_t )? g_t_i : min_g_t;
        }else{ // check for threshold
          if (delta_g_t > 3) {  // TODO: Get a more
accurate value for tGA from experiment
            gyr_exceeded_thr = true;
            }else{
              gyr_exceeded_thr = false;
            }
        }
  }
```

In addition to the fall-detection algorithm implementation, the code needs to generate an alarm signal and send it to the smartphone. The maximum values will be set and reported for accelerometer x-axis as an alarm signal to the smartphone to trigger the process to contact emergency. The following code snippet implements the fall-detection alarm signal:

```
// Check for accumulative
  if ((true == acc_exceeded_thrs) && (true == gyr_ex-
ceeded_thr)) {
     angle = acos(sqrt(accel_x_pwr2 + accel_y_pwr2) /
motionData.accData[2] /* accel_z */) * (180/3.14);
       /* Set max value as a signal to the Smartphone
2^15 - 1 -> max value possible for type int16_t */
       if (angle > 60) // TODO: get a more accurate
value for T_i from experiment
       {
         motionData.accData[0] = ((1 << 15) - 1);
       }
   }
   return;
 }
}
```

10.3.8 Smartphone iOS App

In this project, the smartphone is used as a gateway for the wearable device. It receives the sensors' values, including the signal of a potential fall-detection, and then does two things: Firstly, it sends a programmable text to an emergency contact of choice via SMS if a fall-detection signal is received. Secondly, it logs the acquired data to the Thingspeak cloud platform. Figure 10.31 shows different views of the smart app.

The source code was downloaded from the outdated original Hexiwear Swift version 3 iOS app on GitHub (no need to start from scratch). Then the code was migrated to the newest Swift version at the time this section was written, Swift version 5. Some code was needed to be added since the IoT platform of choice was changed to Thingspeak. The app can post data to the IoT platform and has a view to see a graph of the values versus time. A new module was also added to handle the emergency contact SMS communication.

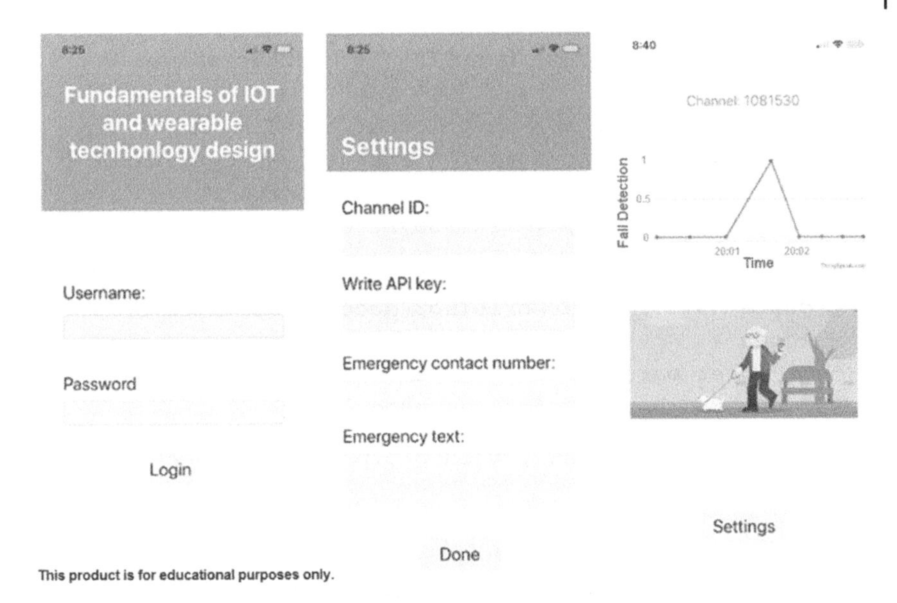

Figure 10.31 Smartphone app different screens.

Once the app is installed, the user can login using the username 'demo' and password 'demo', then they can enter the Thingspeak channel number and the write-API-key so that the smartphone app can post to the cloud IoT platform. The user should also enter the emergency contact number and the SMS text to be sent in case of emergency.

The snippet below shows the Swift code that checks the sensor value and compares it to a threshold to send the emergency text message. The Webkit library is used for reading the cloud channel whereas Alamofire, which is an HTTP networking library written in Swift, is used for sending the SMS text.

```
    var thingSpeakWriteValue: Int = 0
    let accXAlarmThresold = 4.0
    @IBOutlet weak var accountUsernameSIDF: UI-
TextField!
    @IBOutlet weak var authTokenF: UITextField!
    @IBOutlet weak var emergencyNumberF: UITextField!
    @IBOutlet weak var userNumberF: UITextField!
    @IBOutlet weak var textBodyF: UITextField!

    func sendEmergencyText() {
```

```
        var accountUsernameSID = accountUsernameS-
IDF.text!
        var authToken = authTokenF.text!
        var emergencyNumber = emergencyNumberF.text!
        var userNumber = userNumberF.text!
        var textBody = textBodyF.text!

        let url = #"https://api.twilio.com/2010-04-01/
Accounts/\#(String(describing: accountUsernameSID))/
Messages"#
        let parameters = ["From": emergencyNumber,
"To": userNumber, "Body": textBody]

        AF.request(url, method: .post, parameters:
parameters)
            .authenticate(username: accountUsername-
SID, password: authToken)
            .responseJSON { response in
                debugPrint(response)
        }
        RunLoop.main.run()
    }
    func checkValueThreshold(accX: Double) {
       if accX > accXAlarmThresold {  // accXAlarmThresold = 4.0
            sendEmergencyText()
            thingSpeakWriteValue = 1
        }
    }

    @IBOutlet weak var field: UITextField!
    @IBOutlet weak var apiKey: UITextField!

    func postDataToCloud() {
        print(#"https://api.thingspeak.com/update?api_
key=\#(String(describing: apiKey.text!))&field\#(String(des
cribing: field.text!))=\#(String(thingSpeakWriteValue))"#)

       let url = URL(string: #"https://api.thingspeak.
com/update?api_key=\#(String(describing: apiKey.text!)
)&field\#(String(describing: field.text!))=\#(String(t
hingSpeakWriteValue))"#)!
```

```
        let task = URLSession.shared.dataTask(with:
url) {(data, response, error) in
            guard let data = data else { return }
            print("The response is : ",String(data: data,
encoding: .utf8)!)
            //print(NSString(data: data, encoding:
String.Encoding.utf8.rawValue) as Any)
        }
        task.resume()
    }
```

The following code snippet shows the Swift code that displays a field from a Thingspeak channel:

```
import UIKit
import Foundation
import WebKit

internal class ShowLinkController: UIViewController {

    // MARK: - fileprivate Properties -
    // UI
    internal lazy var linkView: ChannelLinkView = {
        let view = ChannelLinkView()
        view.delegate = self
        return view
    }()

    override func viewDidLoad() {
        super.viewDidLoad()
        setupViews()
    }
}

// MARK: - fileprivate Setup Helper Functions -
fileprivate extension ShowLinkController {
    func setupViews() {
        setupMainView()
    }

    func setupMainView() {
```

```
        linkView.translatesAutoresizingMaskIntoCon-
straints = false
        linkView.widthAnchor.
constraint(equalToConstant: 200).isActive = true
        view.addSubview(linkView)
        linkView.topAnchor.constraint(equalTo: view.
topAnchor, constant: 40).isActive = true
        linkView.centerXAnchor.constraint(equalTo:
view.centerXAnchor).isActive = true
    }
}

extension ShowLinkController: ChannelLinkViewD-
elegate {
    func userDidTapOnCannelLink(from: ChannelLinkView) {
        let linkFullView = UIView()
        view.addSubview(linkFullView)
        linkFullView.translatesAutoresizingMaskInto-
Constraints = false
        linkFullView.heightAnchor.
constraint(equalToConstant: 230).isActive = true
        linkFullView.widthAnchor.
constraint(equalToConstant: UIScreen.main.bounds.width
- 17).isActive = true
        linkFullView.topAnchor.constraint(equalTo:
linkView.bottomAnchor, constant: 20).isActive = true
        linkFullView.centerXAnchor.constraint(equalTo:
view.centerXAnchor).isActive = true

        let jscript = "var meta = document.
createElement('meta'); meta.setAttribute('name',
'viewport'); meta.setAttribute('content', 'width=420',
'height=300'); document.getElementsByTagName('head')
[0].appendChild(meta);"
        let userScript = WKUserScript(source: jscript,
injectionTime: .atDocumentEnd, forMainFrameOnly: true)
        let wkUController = WKUserContentController()
        wkUController.addUserScript(userScript)
        let wkWebConfig = WKWebViewConfiguration()
        wkWebConfig.userContentController = wkU-
Controller
```

```
        let webView = WKWebView(frame: linkFullView.
bounds, configuration: wkWebConfig)
        webView.autoresizingMask = [.flexibleWidth,
.flexibleHeight]
        linkFullView.addSubview(webView)
        webView.allowsBackForwardNavigationGestures = true
        let myURL = URL(string: "https://thingspeak.
com/channels/1081530/charts/2?bgcolor=%23ffffff&color
=%23d62020&dynamic=true&results=60&title=Fall+detectio
n&type=line")
        let myRequest = URLRequest(url: myURL!)
        webView.load(myRequest)
  }
}
```

The user can get the app from the Appstore with the name "Hexifall" or get the updated source code from the GitHub link provided in the "References" section.

10.3.9 Cloud Solution

A cloud IoT platform is used to receive and store fall events data, visualize that data, and potentially control those devices and/or react to certain triggers by performing a certain action.

As shown in Chapter 7, there are several cloud platforms for the IoT, and they come with different capabilities when it comes to cost, scalability, ease of use, connectivity, and other features. Some of the best IoT cloud platforms include the Microsoft Azure IoT Suite, Google Cloud IoT Platform, IBM Watson IoT Platform, Amazon AWS IoT Platform, Cisco IoT Cloud Platform, Thingspeak IoT Platform, Oracle IoT Platform, and many more.

The Thingspeak IoT platform is chosen for this example project. It is an open-source platform that provides a way for users to log IoT/wearable device readings on the cloud and visualize the collected data in the form of graphs or charts or other plugins. It can also execute MATLAB® code in order to perform online analysis and processing of the data as it comes in. ThingSpeak is often used for prototyping and proof of concept IoT systems that require analytics. It also has the capability to collaborate with web services, social media networks, and other APIs.

It has a relatively simple user interface, and a RESTful API for the IoT device (or a smartphone in this case) to send data to it.

To send data to a Thingspeak channel, the user needs to sign-up and create a channel then define the fields that will log the data. Once the channel is created,

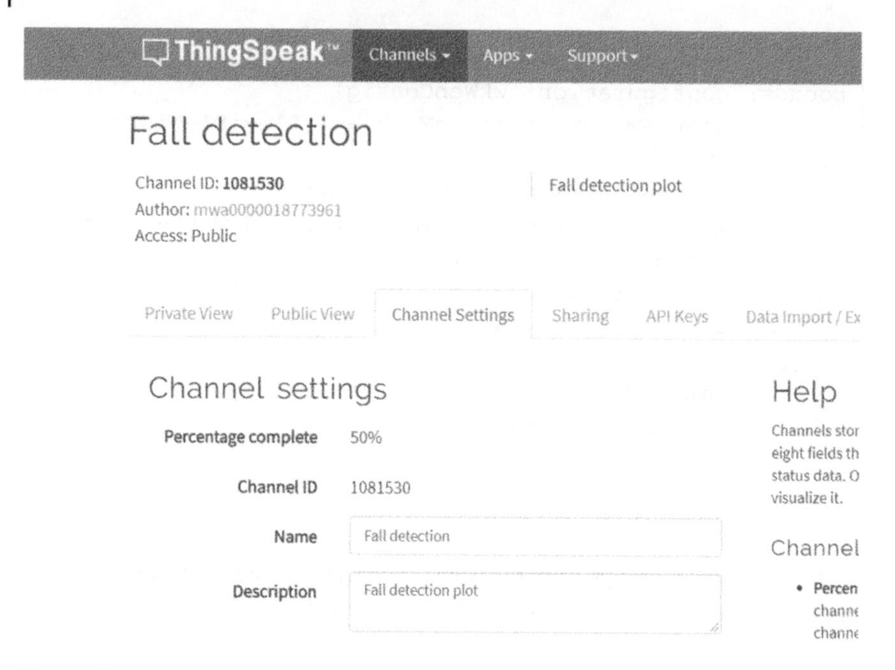

Figure 10.32 Thingspeak channel configuration. *Source:* The MathWorks, Inc.

the platform will provide you with the channel ID to read the channel data, and a write-API-key for the device to write data to the channel. Note that a read-API-key will be needed to read the private fields of a channel. A link to a walkthrough to get started with Thingspeak is provided in the "References" section.

Figures 10.32 and 10.33 show snapshots of the configuration of a typical Thingspeak channel:

10.3.9.1 Cloud versus Edge Computing

As mentioned in Section 3.2.1.3 of Chapter 3, cloud computing does most of the processing at a centralized location. On the other hand, edge computing architecture, or simply edge architecture, is where most of the processing takes place as close to the IoT/wearable device as possible (i.e. close to the data source). Fog computing is the standard that defines the details of the edge architecture. Edge computing has the advantage of limiting the system load which helps in scaling up the number of devices. It is also used for time-sensitive applications since it lowers the system latency by reducing the time it takes for the data to travel to the node where it gets processed. In this project, a hybrid solution is used, edge computing processes the data as soon as possible to trigger an action, and the cloud solution is added for logging purposes with the potential to extend or modify the system capabilities as requirements change.

Channel Stats

Created: about 6 hours ago
Last entry: 5 minutes ago
Entries: 8

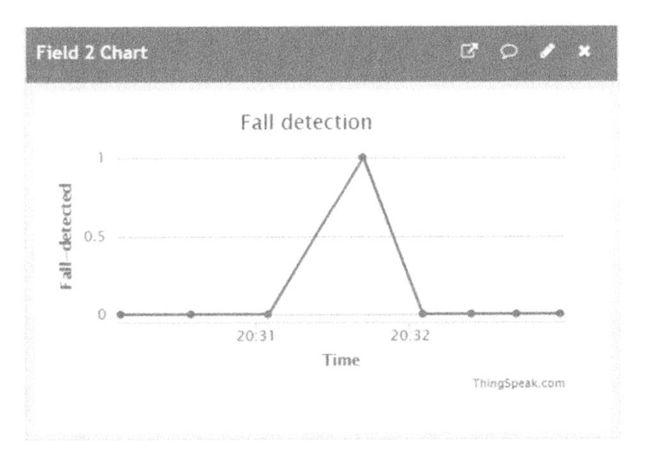

Figure 10.33 Thingspeak channel visualization.

10.3.10 Security

Hexiwear uses BLE to communicate data to the smartphone; hence, all BLE security features are inherited by that interface. Data within the payload is encrypted with the AES-128 block cipher to ensure confidentiality. The smartphone, which acts as a gateway to the internet, uses the restful HTTP API mentioned in Section 10.2.6.2 (RESTful Web Services) to post to the cloud. Additionally, the packets are secured by the SSL/TLS security layer to provide authentication and data integrity which are two of the information security pillars.

10.3.11 Power Consumption

The device's power consumption can be determined using NXP MCU Power Estimation tool as shown in Figure 10.34. Entering the battery information from the Li-pol battery datasheet and enabling the UART and I2C peripherals give a maximum operation time of 15 hours and five minutes, which is longer than the time specified in the requirement section.

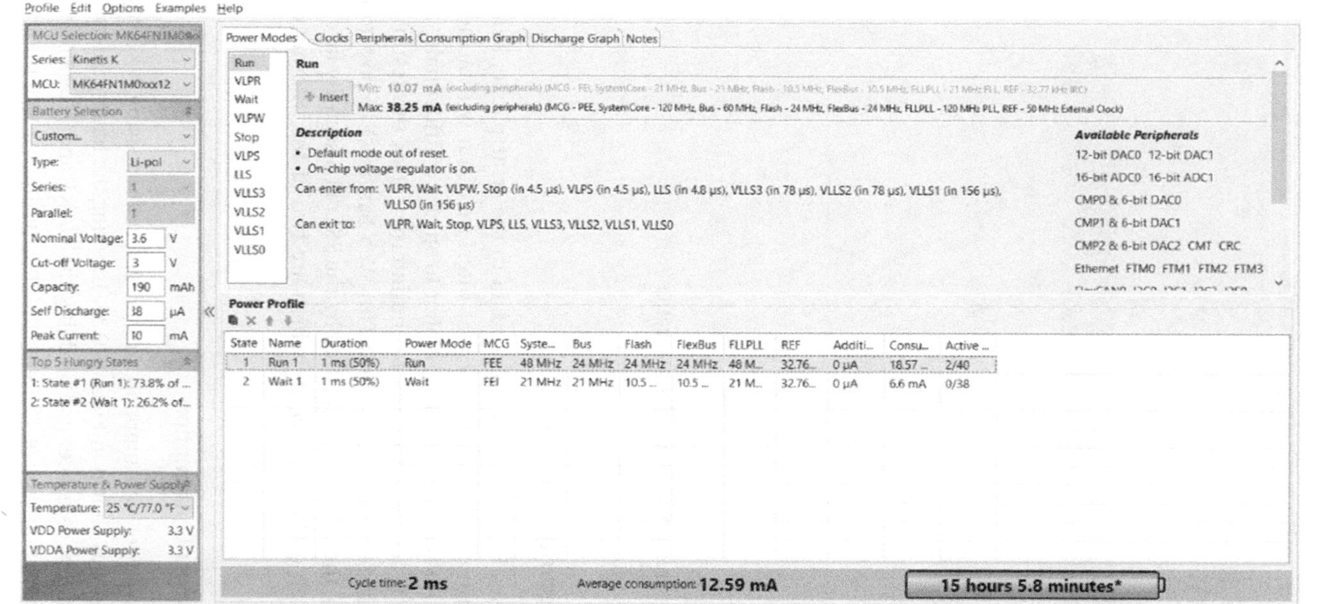

Figure 10.34 Battery life estimation using NXP MCU power estimation tool. *Source:* NXP Semiconductors.

It should be noted that taking into consideration the Bluetooth Smart application example in the MKW40Z Power Consumption Analysis [22], the power consumption of the K40Z board is very low and can be ignored in the practical application of this prototype.

10.3.12 Delivery

By going through the requirements listed in the requirements section, we can see that the prototype satisfies all the criteria:

1) **Accuracy**: The system can reliably detect a person falling (up to 100% in the walking scenario).
2) **Usability**: The system can easily be configured using the user-friendly smartphone app.
3) **Size**: The device is reasonably small, around 1.9″ × 1.75″ × 0.5″.
4) **Weight**: The device is reasonably light in weight, it weighs 40 g.
5) **Power**: The device can maintain operation for at least eight hours.

10.4 Conclusion

Hopefully, during the course of this Chapter, the reader has gained a good understanding of the process of moving from a concept for an IoT/wearable electronic device, through defining requirements, design, PCB printing, and soldering and all the way through to the software and cloud implementation. There are many further resources available to the reader on this subject, and the code and hardware used in this Chapter are all documented, please see the [23, 24] links in the References section.

References

1 DS18B (2019). DS18B datasheet. https://datasheets.maximintegrated.com/en/ds/DS18B20.pdf (accessed June 2020).
2 Fritzing (2010). Fritzing tutorials. http://fritzing.org/learning/tutorials (accessed May 2020).
3 NORDIC Semiconductor Power profiler kit. https://www.nordicsemi.com/eng/Products/Power-Profiler-Kit (accessed May 2020).
4 Boy, T.G. (2016). Arduino CC DS18B20 digital temperature sensor. https://create.arduino.cc/projecthub/TheGadgetBoy/ds18b20-digital-temperature-sensor-and-arduino-9cc806 (accessed May 2020).

5 RocketScream (2018). Rocketscream blog. http://www.rocketscream.com/blog/ (accessed May 2020).
6 RocketScream (2018). Low power lab. https://github.com/LowPowerLab/LowPower (accessed May 2020).
7 Home Automation Community (2020). Atmega328p low power guide. http://www.home-automation-community.com/arduino-low-power-how-to-run-atmega328p-for-a-year-on-coin-cell-battery/ (accessed May 2020).
8 Rohner, A. (2015). How to modify Arduino Pro Mino for low power. https://andreasrohner.at/posts/Electronics/How-to-modify-an-Arduino-Pro-Mini-clone-for-low-power-consumption/ (accessed May 2020).
9 Codeplea (2009). Optimal bit packing. https://codeplea.com/optimal-bit-packing (accessed January 2020).
10 Coranac (2008). Working with bits and bitfields. http://www.coranac.com/documents/working-with-bits-and-bitfields/ (accessed May 2020).
11 IFTTT IFTTT documentation. https://platform.ifttt.com/docs (accessed May 2020).
12 Autodesk Autodesk Eagle overview. https://www.autodesk.com/products/eagle/overview (accessed May 2020).
13 SparkFun (2009). Eagle installation guide. https://learn.sparkfun.com/tutorials/how-to-install-and-setup-eagle (accessed May 2020).
14 SparkFun (2009). Using Eagle schematics. https://learn.sparkfun.com/tutorials/using-eagle-schematic (accessed May 2020).
15 Sprakfun (2009). Using Eagle board layout. https://learn.sparkfun.com/tutorials/using-eagle-board-layout (accessed May 2020).
16 EEVBlog (2011). SMC soldering. https://www.youtube.com/watch?v=b9FC9fAlfQE (accessed March 2020).
17 Adafruit (2017). Adafruit BME280 library. https://github.com/adafruit/Adafruit_BME280_Library (accessed March 2020).
18 Sparkfun (2014). Soil moisture sensor hookup guide. https://learn.sparkfun.com/tutorials/soil-moisture-sensor-hookup-guide (accessed March 2020).
19 Sparkfun (2009). Generating gerber files. https://learn.sparkfun.com/tutorials/using-eagle-board-layout#generating-gerbers (accessed May 2020).
20 Hexiwear (2016). Hexiwear user manual. https://www.mikroe.com/blog/hexiwear-user-manual (accessed June 2020).
21 MikroElektronika/HEXIWEAR (2017). Hexiwear stock firmware source code. https://github.com/MikroElektronika/HEXIWEAR/tree/master/SW/MK64%20KDS (accessed June 2020).
22 Rakhman, A.Z., Nugroho, L.E., Widyawan, W., and Kurnianingsih, K. (2014). Fall detection system using accelerometer and gyroscope based on smartphone. *2014 The 1st International Conference on Information Technology, Computer, and Electrical Engineering*, Semarang, pp. 99–104.

23 GitHib (2020). GitHib link. https://github.com/mkamoona/FundametnalsOfIoT (accessed May 2020).

24 Github (2020). All source code and Eagle schematics for this chapter are available here. https://github.com/statts/siguino (accessed March 2020).

Further Reading

Valvano, JW (2014) *Embedded Systems: Introduction to ARM®CORTEX-M Microcontroller*, 5e, vol 1 Texas, US: Valvano.

Freescale Semiconductor (2015a) FXOS8700CQ 6-axis combo accelerometer and magnetometer datasheet https://osmbedcom/media/uploads/GregC/fxos8700cq_ds_rev6pdf (accessed March 2020)

Freescale Semiconductor (2015b) FXAS21002CQ 3-axis gyroscope datasheet https://osmbedcom/media/uploads/GregC/nxp_fxas21002cq_datasheetpdf (accessed March 2020)

Freescale Semiconductor, Inc (2015) Kinetis K64F MCU data sheet https://osmbedcom/media/uploads/GregC/k64f_ds_rev6pdf (accessed March 2020)

MikroElektronika Li-Pol battery datasheet https://downloadmikroecom/documents/datasheets/HPL402323-2Cpdf (accessed May 2020)

MikroElektronika/HEXIWEAR (2017) Hexiwear stock firmware binary https://githubcom/MikroElektronika/HEXIWEAR/tree/master/SW/MK64%20KDS/HEXIWEAR_MK64/binary (accessed May 2020)

Multi-cb Multi-CB index https://wwwmulti-circuit-boardseu/en/indexhtml (accessed May 2020)

NXP Semiconductors (2016) MKW40Z power consumption analysis https://wwwnxpcom/docs/en/application-note/AN5272pdf (accessed March 2020)

Thingspeak (2014a) Walkthrough to get started with Thingspeak https://wwwcodeprojectcom/Articles/845538/An-Introduction-to-ThingSpeak?fbclid=IwAR1l6_as4F4g1FxtxfHRhZ-yhrELW73Ul8ieiz0jLDnHiU35e4B0EQzqT-o (accessed May 2020)

Thingspeak (2014b) Getting started with Thingspeak (cloud solution) https://wwwcodeprojectcom/Articles/845538/An-Introduction-to-ThingSpeak?fbclid=IwAR1l6_as4F4g1FxtxfHRhZ-yhrELW73Ul8ieiz0jLDnHiU35e4B0EQzqT-o (accessed May 2020)

Index

Fundamentals of IoT and Wearable Technology Design, First Edition. Haider Raad.

Published 2021 by John Wiley & Sons, Inc.

d

e

f

g

h

Solution Manual

Chapter 1 Homework Problems:

1 What are the main differences between IoT and Wearable Technology?
  A Communication is IP based in IoT, while it's not necessary in wearables.
  B Most wearables rely on a gateway device, such as a smartphone, for configuration and connectivity, and in most cases to enable features and process data. This is not always true in IoT devices.
  C IoT devices are mainly stationary, wearables on the other hand are mobile since they are worn/ integrated within the user's body or clothing.

2 What is it meant by "things" in Internet of Things?
The core functionality of IoT and wearable devices starts with data acquired or an action performed by a device. These devices are called endpoints, and they are the "Things" in Internet of Things. The value of IoT and wearable devices is in the data collected by these endpoints, so it is important to understand how they acquire, process, transmit, and receive data.

3 What are the main differences between IoT and M2M?
  A Communication is IP based in IoT, while it's usually not in M2M.
  B M2M is mainly point to point while this is not true when it comes IoT.
  C M2M devices are stationary, IoT could be stationary, or portable/mobile.

4 Can you think of other potential challenges found in IoT and wearable technology other than the ones mentioned in this chapter?
  A There are several other challenges besides the ones mentioned in the chapter which include: Design based challenges, safety, longevity, compatibility, etc.

Fundamentals of IoT and Wearable Technology Design, First Edition. Haider Raad.

Published 2021 by John Wiley & Sons, Inc.

5 Give examples of wearable devices/applications that do not require internet connectivity.
Ultraviolet exposure wearable device, a simple pedometer, smart socks, GPS enabled hiking helmets, etc.

6 List five real world examples of smart clothing.

A Smart fashion applications (e.g.: Tommy Jeans Xplore) which utilizes an integrated chip that can track how often the product is used and also where it was worn.

B Smart yoga pants (e.g.: Nadi X) which can sense when yoga poses need adjustment by using haptic feedback to create small vibrations on the body part.

C Athlete recovery applications (e.g.: Under Armour's apparel) that absorb heat from the user's body and reflects it back in the form of far infrared light, which is supposed to promote muscle recovery.

D Smart Fitness Socks (e.g.: Sensoria) which use advanced textile sensors to provide precise data on how your foot lands while walking or running.

E A swimming suit equipped with a UV sensor (e.g.: Neviano's swimsuits). The sensor is waterproof and connects to the wearer's phone to send alerts when UV levels are high and more sunscreen should be applied.

7 List five real world examples of the headwear form in wearable technology.
Virtual reality headsets (e.g.: Oculus), Smart motorcycle helmets (e.g.: Sena), Smart ski goggles (e.g.: RideOn), Smart hats (e.g.: LifeBeam), Smart sleep headbands (e.g.: Philips).

8 List four components common between IoT and wearable devices (an application of your choice).
Microcontroller, sensor, battery/power management system, LCD screen.

9 Are wearable devices a form of M2M? Why?
Typically, wearables are non-IP based, and this feature is common with M2M. Although some IoT devices do not directly utilize IP, the data traffic of the networks involved are typically based on IP.

10 If you are asked to add more somewhat essential characteristics to IoT, what would they be? Why?
Cost effectiveness and energy-efficiency are two additional characteristics can be added to the ones listed in the chapter. If these devices did not provide

the potential of an immense value at a low cost, there wouldn't be discussions about developing solutions based on these technologies in the first place.

Chapter 2 Homework Problems:

1 Can you think of more applications (other than the ones listed in this chapter) that could benefit from IoT and wearables?
Restaurants and cooking, weather and natural disasters, waste management, etc.

2 Create a novel scenario where drivers and/or pedestrians could benefit from IoV.
One possible scenario is tailgating detection which can be determined from proximity sensors, street speed limit, and cars velocity.

3 Create a novel scenario where governments could benefit from IoV.
One possible scenario is to tracking and locating criminals. Further, according to one study, it was estimated that deploying IoV can result in $178.8 billion in societal benefits annually in the US.

4 Create a scenario where home automation is utilized in the field of safety.
One possible application is smart fire and smoke detector (IoT based) which allow remote monitoring and improved alerting system.

5 Could you think of more potential applications of IIoT?
Self-driving tractors, smart mining, connected manufacturing, oil field innovations, etc.

6 List five unusual applications where IoT and wearables are utilized. Keep efficiency and practicality in mind, and make sure that no products exist that support such applications (through an internet search).
This is an open-ended question that requires creative thinking.

7 List ten applications where wearables are used in healthcare.
Personal EKG, Alzheimer, Remote Patient Monitoring, Pregnancy Parameters, Smart Hearing Aids, Chronic Pain Management, Chest Band (Respiratory and Heartrate), Arrhythmia Detectors, Posture Correction, Diabetes.

8 Write a one page scenario where at least ten of the applications mentioned in this chapter are utilized in a typical day.
This is an open-ended question that requires creative thinking.

Chapter 3 Homework Problems:

1 Why an architecture is needed for connected devices (IoT and wearables)?

While some similarities between IT and connected devices network architectures do exist, in most cases, the challenges and requirements of IoT and wearable systems greatly differ from those of conventional IT networks.

IT networks are essentially concerned with the infrastructure that transports data, regardless of its type. The main goal of IT networks is the reliable and uninterrupted support of enterprise applications such as email, websites, and databases. On the other hand, networks of connected devices are about the data generated by sensors and how it is used. Thus, the core of such architectures is about how the data is transported, aggregated, processed, and eventually acted upon. Hence, a new architecture is needed.

2 What is the difference between centralized and de-centralized networks?

In general, a decentralized network architecture distributes workloads among several entities, instead of relying on a single entity such as a central server. This trend is enabled thanks to the rapid improvements in the computational power of microprocessors which now offer a performance well beyond the needs of most applications of connected devices.

3 List three published IoT architectures and research three more from the literature. Compare the six architectures using a table.

This is an open ended question which requires the reader to research the literature.

4 Give an example of an IoT device and explain its operation using the simplified IoT architecture reported in this chapter.

This is an open ended question which requires the reader to research the literature.

5 What is Edge Computing? Give four examples of IoT and wearable devices and explain their operation within the context of Edge.

Edge Computing describes the work that happens at the edges of the IoT network, where the physical devices connect to the cloud, exploiting mobile phones, smart devices, and/or network gateways to perform tasks and provide services on behalf of the cloud. With an emphasis on reducing latency, improving privacy and security, and minimizing bandwidth costs within data-driven IoT applications, edge computing architectures are becoming increasingly common in the realm of IoT and wearable devices.

The aim of Edge Computing is to bring computing, and data filtering and storage closer to the devices where it's being collected, rather than relying on a central site that can be thousands of miles away. This is done so that data does not suffer from latency issues that can affect an application's performance. Moreover, enterprises can save money by having the processing performed locally, reducing the amount of data that needs to be processed at the cloud.

A self-driving car generates roughly ten gigabytes of data per mile. If self-driving vehicles on autopilot continue growing in number, it will be impossible to send data to centralized servers for processing every time a vehicle encounters a stop sign or a pedestrian. A microsecond of time is of significant importance in such scenarios. Here is where edge computing comes into play. Other examples include predictive maintenance, voice home assistants (Alexa, Google, etc.), and oil and gas industry.

6 What is the difference between Cloud, Fog, and Mist? Explain using two practical examples (one IoT device, and one wearable device).
Cloud, Fog and Mist computing are all different because they all compute and analyze data inputs at different points within the network and, as a result, all have different latencies and calculating powers. Cloud refers to a large, centralized data center that can make calculations and store data, but is a significant distance from the devices. Fog refers to smaller nodes that are at the edge of the network that can make simpler calculations without needing to send it into the cloud. Mist refers to applications within the device itself that is able to make basic, low-level calculations. For instance, IoV accident prevention feature, or a smart armband that measures blood sugar level of a diabetic wearer would likely use either Fog or Mist computing so it can alert the wearer or another device if an issue is occurring much sooner. Whereas, a smart home system might be more suited to have support on a Cloud system for its processing, even with the slight latency.

7 What are the main differences between IT and IoT networks?
IT networks are essentially concerned with the infrastructure that transports data, regardless of its type. The main goal of IT networks is the reliable and uninterrupted support of enterprise applications such as email, websites, and

databases. On the other hand, networks of connected devices are about the data generated by sensors and how it is used. Thus, the core of such architectures is about how the data is transported, aggregated, processed, and eventually acted upon.

8 How is the OSI model related to IoT and wearable technology architectures?
The OSI model relates to IoT and wearable technology architectures because these architectures are often similar in structure and functionality to the OSI model. These models are all layered in structure, with each layer encompassing various parts of the IoT and wearable technology's functionalities, from the applications and sensors to their connections to the network.

9 Design a basic wearable fitness tracker using the wearables architecture described in this chapter.

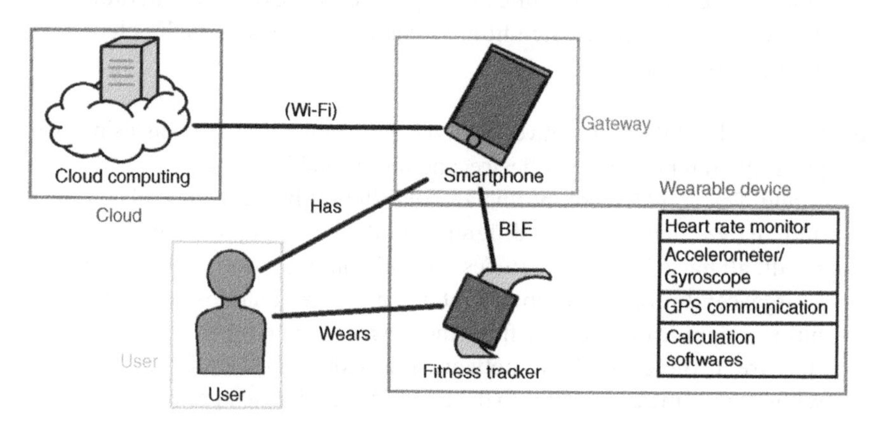

10 Design a basic IoT garden monitor using the simplified architecture described in this chapter.

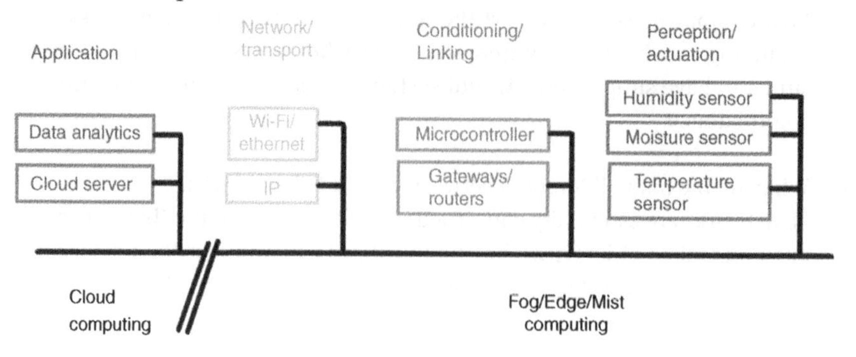

11 Sketch a smart home system and link each component that you use (software and hardware) to an architecture of your choice.

Using the basic architecture:

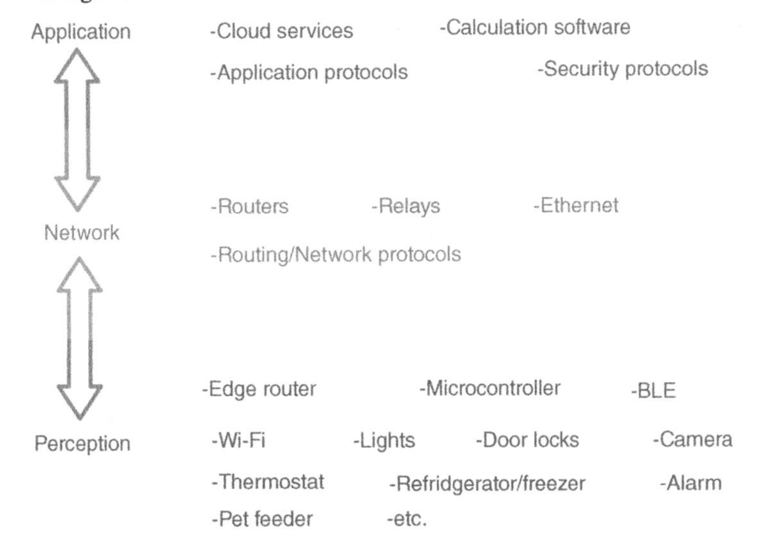

Chapter 4 Homework Problems

1 Based on the anatomy of a general connected device depicted in Figure 1, sketch a similar diagram pertaining a smart watch.

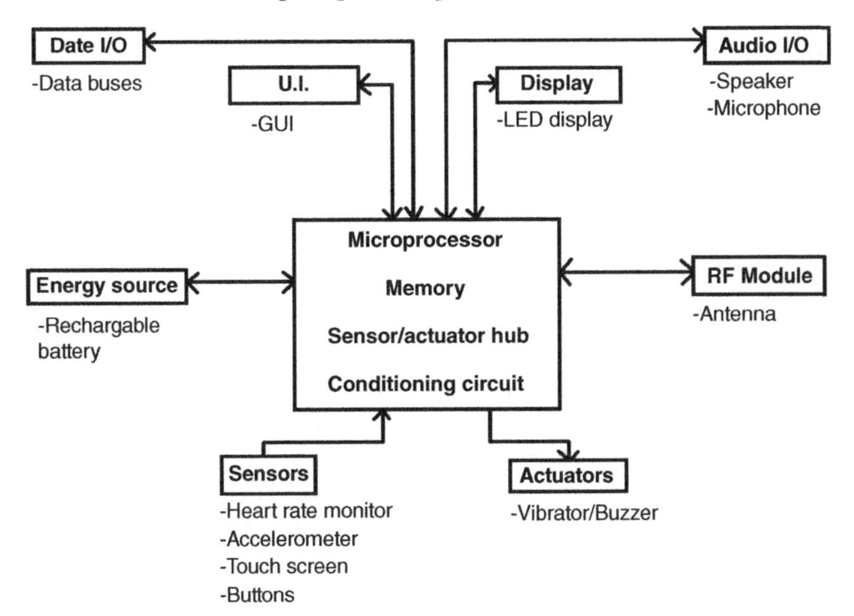

2 Based on the anatomy of a general connected device depicted in Figure 1, sketch a similar diagram pertaining a smart camera-based smart lock and doorbell system.

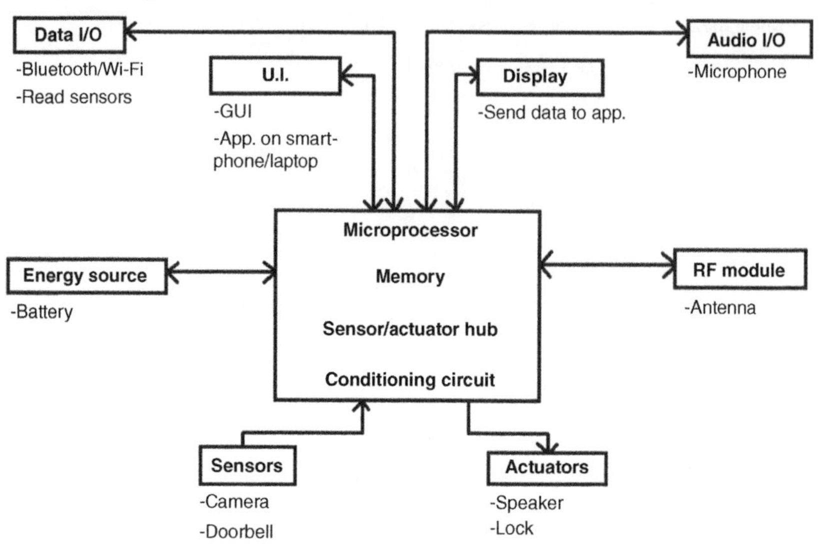

3 Pick a wearable or IoT device of your choice, then list all of the device's components (external and internal).

For a soil moisture sensor, the components would be:

- Temperature Sensor
- Moisture Sensor
- Humidity Sensor
- Pressure Sensor
- LED lights
- Outer casing
- Microcontroller
- Power Source/Battery
- BLE/Wi-Fi-Connective Unit

4 What is a MEMS sensor? Research 5 examples from the literature and compare between their mechanisms of operation.

A MEMS sensor is a short term for micro-electromechanical sensor. One example is an airbag sensor for a vehicle, which uses an accelerometer to sense when great deceleration occurs, thus opening the airbags. A second example is a disposable blood pressure sensor, which uses piezo-resistive material to convert stress into electrical signals that reflect the current pressure in the user's blood. A third example is an inkjet printer head, which uses resistive material that is given electrical signals to heats up ink to form bubbles to push it out of a nozzle. A fourth example is an overhead projection display, which uses many small rotatable mirrors that reflect light and guide it through a lens to produce sharper images on a projection screen. One last example is RF MEMS, which uses RF technologies to obtain and distribute data.

5 What are the different types of accelerometers? How would you characterize a typical one?
Some different types of accelerometers include compression accelerometers, which applies a force to a crystal on a mass-sensing sensor to indicate the amount and direction of motion; shear accelerometers, which uses analyzes shear stress on crystals using piezoelectric sensors to determine the motion; and capacitive accelerometers, which senses the change in capacitance when the capacitor undergoes acceleration. Typically, an accelerometer is a sensor that is able to detect changes in orientation and/or acceleration by an external force and produces electrical signals according to these changes.

6 What are some of the most common types of thermocouples?
Type J Thermocouple: comprises an iron and a copper-nickel alloy leg and is considered the most common thermocouple in use in the US. Type K and N Thermocouples are other types that are known to be accurate and stable at high temperatures, while Type C is used in extremely hot environments. Type T Thermocouple is for cold environment applications.

7 Research the most common types of motors used in IoT and wearable applications.
Different types of motors can be used depending on the application. Servo motors, stepper motors, DC motors are all used.

8 What would be a good choice of an antenna topology for a fitness tracker? Why?
Chip, PCB, and wire antennas are widely used in wearables due to their small form factor. Miniaturization techniques (such as folding and winding) are usually used to save space.

9 What would be a good microprocessor/microcontroller choice for a wearable device that makes one heart rate reading every 6 hours? Justify your choice.
One of the ultra low power processors such as ARM Cortex based (i.e. Cortex®-M0+, M3, M33, and M4) would be sufficient. Eight bit would work since this application is not data-heavy.

10 You are tasked to prototype a virtual home assistant (i.e.: similar to Amazon Echo Dot). Make a list of all the tasks needed to create such a device, along with a list of all the components needed based on what you have learned in this chapter.

For the virtual home assistant, the tasks needed are:

- Determine the specific sensors, actuators, battery and microcontroller to use for the device
- Allow it to connect to a cloud server to send and retrieve data
- Allow secure and safe connections between it and the cloud server
- Allow it to be able to comprehend and analyze the voice samples given to it on the device
- Create and install software to comprehend the voice samples

The components needed are:

- Microphone
- Speaker
- LEDs
- Buttons
- Battery/Power Source
- Microcontroller
- Wi-Fi Connective Unit

Chapter 5 Homework Problems

1 You have narrowed down your choice for a network topology to either a full mesh topology or a star topology. Determine how your final decision will affect deployment cost and communication speed.

If one was to go with a full mesh topology, then communication speed will be faster because of the direct connections between each of the nodes on the network, whereas with a star topology, any data sent between end systems would be sent through the central node before reaching its destination. However, the cost would probably be lower for the star topology because there are overall less connections within the entire network compared to the full mesh.

2 Based on literature research, comment on how IP addresses are arranged and displayed.

IP addresses are arranged and displayed usually as four decimal numbers separated by periods. In reality, they are a 32-bit binary number. The number is split into four bytes made up of eight bits each and displays the decimal equivalent of the binary number that makes up the byte, to make it easier to read and comprehend quickly. As a result, each of the four numbers is within the range of 0 (00000000_2) to 255 (11111111_2). A comparison between IPv4 and IPv6 headers can be found here: https://www.cisco.com/en/US/technologies/tk648/tk872/technologies_white_paper0900aecd8054d37d.html

3 You are designing a fitness tracker. What would your protocol and topology choices be?
For a fitness tracker, one would use protocols such as 6LoWPAN-UDP-DDS and a Point to Point topology. The fitness tracker would likely only transmit its data to one other device, such as a cell phone or laptop, so the Point to Point topology would be used in conjunction with the DDS while UDP and 6LoWPAN would provide reliable, low-power data transfer between the two devices.

4 Sketch a protocol stack for a smart IoT-based thermostat.

Application	DDS
Transport	UDP
Network	6LoWPAN
Physical/Link	Wi-Fi

5 Sketch a protocol stack for a smartwatch. Compare the flow of data with the one you sketched in the previous question.

Application	MQTT
Transport	TCP
Network	IPv6
Physical/Link	Cellular Technology

Compared to an IoT-based thermostat, a smartwatch likely has much more features including the ability to read and send text messages. Therefore, more wide-range and faster data transmission is likely that can connect to multiple devices what is going to be used as opposed to the thermostat which likely connects to only one device. It must also have reliable data transmission.

6 Explain why it is better to use 6LoWPAN-UDP-CoAP stack in IoT instead of IPv6-TCP-HTTP.
The 6LoWPAN-UDP-CoAP stack is better to use because it is more reliable than the IPv6-TCP-HTTP stack. This is because the former uses UDP which, while slower than TCP, features more reliable and lossless data transfer between devices. Furthermore, both 6LoWPAN and CoAP are better for devices that are either restrained by the amount of power they can use of the amount of resources they have respectively.

7 A network with all the nodes acting as both servers and clients. A PC can access files located on another PC but also delivers files to other PCs on the network. Which network architecture is that?
This describes a mesh architecture because each PC has direct connections to every other PC on the network.

8 Which of the following is NOT an advantage of a star network topography?
A There is no central point of failure
B Easy to add or remove a node as it has no effect on any other node
C Reasonable security (i.e.: no node can interact with another without passing through the server first)
D Few data collisions as each node has its own connection to the server

The answer is (a) There is no central point of failure. In a star topography, if the node that acts as the central hub fails, then its connections to all of the other nodes in the topography are lost, meaning the entire network fails. The hub node is the central point of failure.

9 Which layer does the Ethernet and Wi-Fi protocols belong to?
Ethernet and Wi-Fi protocols belong to the Physical/Data Link Layer.

10 What happens to the packet as it is passed from the application layer to the transport layer? What about from the transport layer to the network layer?
When a packet is passed from the application layer to the transport layer, it gets packaged together and the transport layer adds information such as transport protocol, and information on the packet's source and destination. When a packet is passed from the transport layer to the network layer, that information is packed together and the transport layer adds information such as the network protocol and how it goes about moving through the network to reach its destination.

Chapter 6 Homework Problems

1 A smart thermostat system uses a temperature sensor and a microcontroller with a Bluetooth connectivity. What is the capacity of the battery that you would choose for the device to last at least 6 years? Assume that the processor clock frequency is 50 MHz, communication current is 3.5 mA, data logging

current is 25 μA, sleep mode current is 2 μA, and wakeup time is 140 μs. Communications run for 0.25 s every hour, data logging runs for 20 ms every second.

$$I_{comm} = 3.5\cdot10^{-3}\left[A\right] \quad I_{log} = 2.5\cdot10^{-5}\left[A\right] \quad I_{slp} = 2\cdot10^{-6}\left[A\right]$$

$$t_{comm} = 0.25\left[s\right] \quad t_{log} = 72\left[s\right] \quad t_{slp} = 3527.75\left[s\right] \quad t_{operating} = 52560\left[hr\right]$$

$$I_{av} = \frac{I_{comm}t_{comm} + I_{log}t_{log} + I_{slp}t_{slp}}{t_{comm} + t_{log} + t_{slp}} = \frac{0.0097305\left[A\cdot s\right]}{3600\left[s\right]} = 2.703\cdot10^{-6}\left[A\right]$$

$$Capacity = I_{av}t_{operating} = \mathbf{0.14206}\left[\boldsymbol{Ah}\right]$$

The average current was calculated based on the average per hour.

2 A battery-operated IoT device must run for 2 years without replacing the battery, at Irun =25 mA, with 1 ms operation for every 2 seconds and sleep current Islp=1 μA. Determine the required battery capacity?

$$I_{av} = \frac{I_{slp}t_{slp} + I_{run}t_{run}}{t_{slp} + t_{run}} = \frac{10^{-6}\left[A\right]\cdot1.999\left[s\right] + 0.025\left[A\right]\cdot0.001\left[s\right]}{2\left[s\right]}$$
$$= \frac{0.026999\left[A\cdot s\right]}{2\left[s\right]} = 1.34995\cdot10^{-6}\left[A\right]$$

$$Capacity = I_{av}t_{operating} = 1.34995\cdot10^{-6}\left[A\right]\cdot17520\left[h\right] = \mathbf{0.23651}\left[\boldsymbol{Ah}\right]$$

3 Pick a wearable device of your choice then list three battery candidates available commercially for a reasonable operating time. Justify your battery choice according to the energy budget of the wearable device.
For a wrist-mounted device that measures heart rate, of the battery choices listed below, one would go a battery with a capacity in the range of 150-350 mAH due to its reasonable battery capacity in conjunction with its small physical size, because the device in question is not one that has a lot of room for large batteries. Below are examples of a few candidates:

CR2025 (capacity: 170 mAH)
CR2032 (capacity: 210 mAH)

4 What is the wavelength at 900 MHz, 2.45 GHz, and 60 GHz? What is the path loss over 1 meter, 100 meters, and 1 kilometer for these frequencies? Assume dipole antennas (Gain = 1.7 dB)

900 MHz (Wavelength = 0.33 m), Path loss: 1 m = 28.1 dB, 100 m = 68.1 dB, 1 km = 88.1 dB

2.4 GHz (Wavelength = 0.122 m), Path loss: 1 m = 36.8 dB, 100 m = 76.8 dB, 1 km = 96.8 dB

60 GHz (Wavelength = 0.0049 m), Path loss: 1 m = 64.6 dB, 100 m = 104.6 dB, 1 km = 124.6 dB

5 What is the maximum range of a fitness tracker connected to a smartphone using BLE. Explain in terms of link budget analysis.
The maximum range of a fitness tracker connected to a smartphone using BLE is proportional to the power transmitted by the tracker, the power that is received by the smartphone, the gains of the antennas used and other propagation losses, as given by the Friis equation. These values correspond to a certain distance, and the distance which the Friis equation makes it out of range, thus reaching the range of communication for the devices.

6 A WiMax base station transmits at power levels of 43 dBm, with an antenna gain of 14 dBi, and a receiving sensitivity of -92 dBm. An IoT irrigation system is located two miles away with a dipole antenna of 1.76 dBi gain, a transmitting power of 16 dBm, and a receiving sensitivity of -88 dBm. The cables and connectors have a loss of 3 dB at each end. Is the communication link feasible?

$$D = 2[mi] = 3218[m], \quad P_t = 43[dBm], \quad P_r = 16[dBm], \quad G_t = 14[dBi],$$
$$G_r = 1.76[dBi] \quad L = 3[dB]$$

$$\begin{aligned} 20\log\left(\frac{\lambda}{4\pi}\right) &= P_r + 20logD - P_t - G_r - G_t - 2L \\ &= 16 + 20\log(3218) - 43 - 14 - 1.76 - 6 \\ &= 16 + 70.15 - 43 - 14 - 1.76 - 6 = \mathbf{21.39}[\mathbf{dB}] \end{aligned}$$

The connection is feasible.

7 Sketch a flow diagram for the development process of a basic fitness tracker.

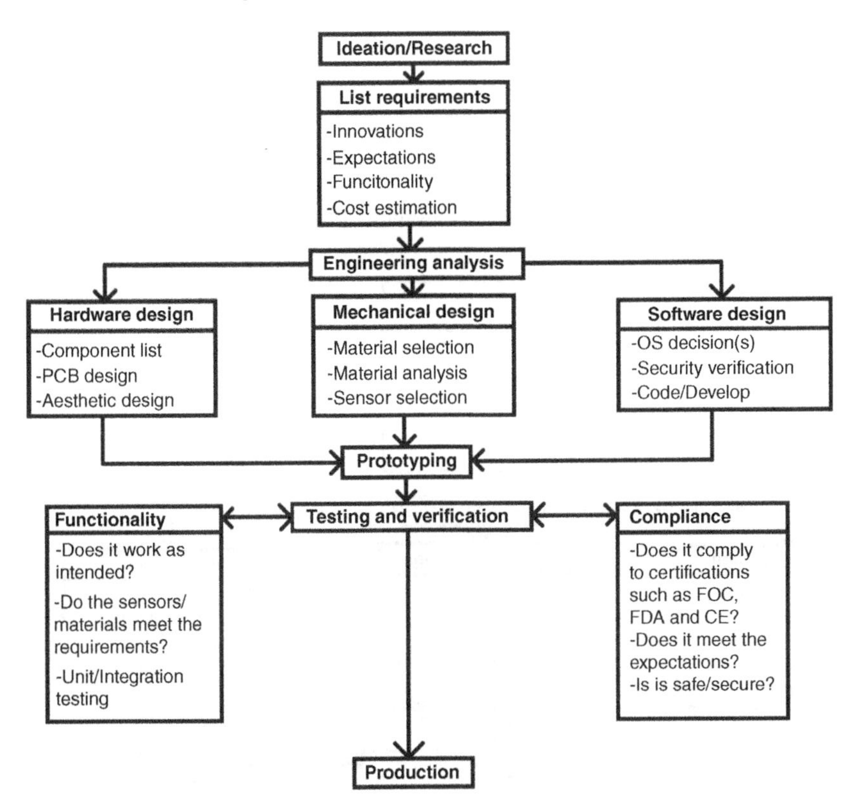

8 List all the possible design considerations you believe they are appropriate for a smart T shirt that measures heart rate, breathing rate, and temperature.

The design considerations for this smart t-shirt include:

- How quickly and how accurately the shirt measures the desired parameters
- How well the circuitry of the shirt can handle the range of temperatures, weather conditions, the wearer's sweat or other bodily fluid, the constant motion of the wearer, and other physical conditions it could be exposed to
- How simple the parameters that are being measured could be accessed or viewed by the wearer or another observer
- Whether or not the shirt will connect to an external device to view or process data
- How comfortable the circuitry feels wearing on the shirt and how it aesthetically alters the design of the shirt itself
- Whether the materials used in either the shirt or the circuitry is comfortable on human skin and easy to move around in

- Whether the circuitry can be cleaned or washed without much degradation
- How long the shirt is expected to function without changing out its battery or power source
- How much memory or computational power the shirt will have
- How can external devices can connect to and view the parameters measured by the shirt

9 Sketch a flow diagram for the development process of an IoT based security system.

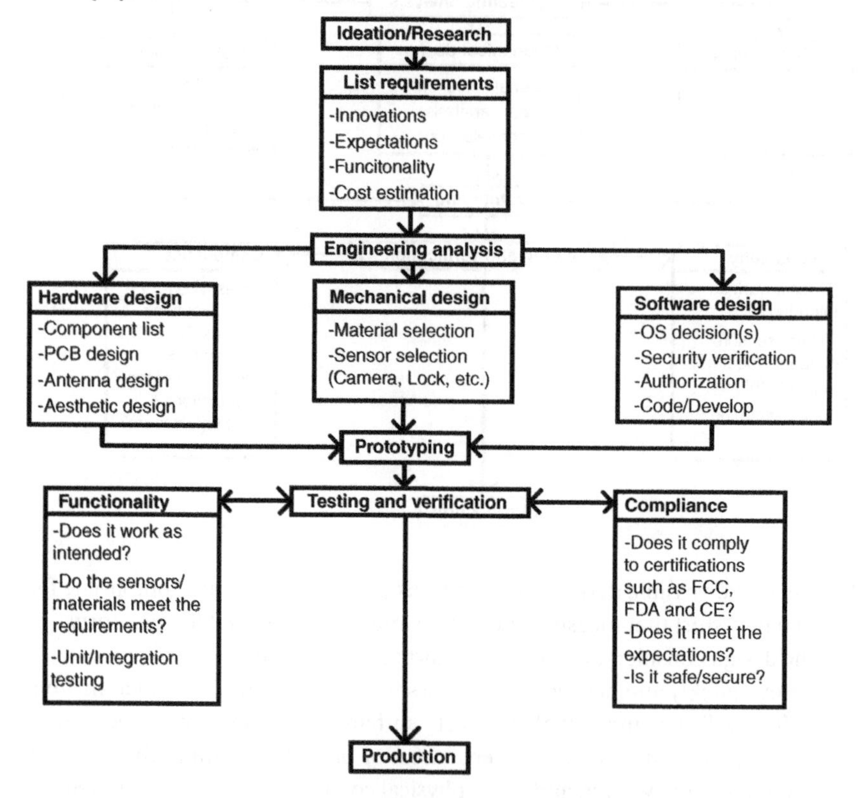

10 List all the possible design considerations you believe appropriate for an automatic dog feeder. The device lets you feed your dog remotely, schedule meals, and control the portion size.
The design considerations for this dog feeder include:

- How reliable and accurate the data sent to the feeder is
- Whether the packaging is durable enough to prevent tampering with the circuitry

- Whether the data and specifications sent to the feeder are reliable and accurate
- How and how easy it is for the user to connect to and use the feeder
- How the external aesthetics look, its total weight and whether they conceal both the food and circuitry
- The connectivity to other compatible devices so that the user can use the feeder remotely
- Whether the feeder is safe to be active when the user or their dog is around, especially around where the user puts the food in and where it is dispensed.
- How easy it is to identify potential problems with the feeder, both physically and with the software used
- How easy it is to connect and authorize users to the feeder and prevent unauthorized users from making changes to the feeder's specifications

Chapter 7 Homework Problems:

1 Research examples of IoT and wearable devices with each example utilizing one of the cloud types mentioned in this chapter.
An example of a private cloud could be an IoT-based smart home system because all of the specific devices and their specific information and parameters could be stored securely with limited access to outside devices. An example of a public cloud could be a streaming service that can deliver data or information that it has stored in the cloud to various devices that are connected to the cloud's network. An example of a hybrid cloud could be a smartwatch, where a lot of its applications and data can be sent to external clouds while the rest of its features are embedded publicly into the device itself. An example of a community cloud could be a smart city that exchanges and stores information between the many different devices and servers to fulfill its features.

2 Research examples of IoT and wearable devices with each example utilizing one of the cloud service models mentioned in this chapter.
An example of IaaS could be the Google Compute Engine. An example of SaaS could be GitHub. An example of PaaS could be Heroku. An example of FaaS could be OpenFaaS. These could be used in conjunction with IoT devices so that they can establish their structure and usage.

3 You are working on developing a new smart home virtual assistant. Research five of the IoT platforms mentioned in this chapter, then narrow down your selection for this project to two candidates. Justify your choices.

IoT Platform	Features and Functionalities
AWS	Provides on-demand cloud computing, APIs, storage of objects, databases, management and developer tools, among other features
Google Cloud IoT	Uses Google's resources to provide tools that work well with supply chain management, location intelligence and predictive maintenance
Azure IoT Suite	Provides solutions to presented problems that updates as the project is customized, allowing for easy prototyping and sampling
Oracle IoT	Provides SaaS applications, end-to-end security, broad support and built-in AI, among other features
Salesforce IoT	Works to monitor and support the devices that a company provides to customers by sending assistance remotely

Of these five IoT platforms, for a smart home system, one good choice would be the AWS IoT platform because it provides support to a broad amount of devices and protocols, and since a smart home would need to support lights, locks, cameras, and other devices, the broadness and the security it provides would be useful for a home.

4 List five applications or devices that could benefit from a fog layer. Justify your answers.

Some applications or devices that could benefit from a fog layer include smart vehicles, because it would need to make quick decisions on keeping passengers and those around it safe; medical monitors, because it cannot delay to alert health care workers or guardians if an issue arises; smart proximity detectors, such as one at an intersection that could quickly alert pedestrians of any oncoming traffic; and security devices, to send an alert as quickly as possible in the event of a break-in.

5 Research five open source API platforms and compare between their functionalities, features, and related criteria in a comparison table.

API Platform	Features and Functionalities
API Umbrella	Rate Limiting, API Keys, Caching, Real-Time Analysis, Admin Permission Varying
Gravitee.io	Rate Limiting, IP Filtering, Cross-Origin Resource Sharing, Developer Portals, good for comprehending data usage
APIman.io	Quick Runtimes, Asynchronous Capability, Policy-based Governance, Billing and Analytic Options
WSO2 API Manager	High Customization, Easy Governing Policies, Better Access Control, Runs anywhere at anytime
Kong Enterprise	Good for management, One-Click Operations, Software Health Checks, Availability of Open-Source Plugins

[Reference] https://appinventiv.com/blog/open-source-api-management-tools/

6 Comment on the mechanism of a simple REST client example for retrieving API data from an IoT or wearable device.
REST uses HTTP calls in order to retrieve the data from an API, whether it is an operation that needs to be run or data regarding other protocols or routines that the IoT or wearable device would need to operate. This means that REST clients are connected to the Web in some way if they are using HTTP calls to access information that is stored in other devices and servers.

7 Pick an IoT or wearable technology application and comment on how involving a machine learning algorithm will lead to more useful insight.
One IoT device that could benefit greatly from machine learning and AI is that of smart vehicles. As the technology used for it develops, the AI can teach itself good driving techniques and form decision-making operations to make quick decisions that result in the safest outcome for the passengers of the vehicle as well as those of other vehicles and nearby pedestrians that it could detect.

8 Compare through a table the differences between the OpenStack and OpenFog architectures.

OpenStack	OpenFog
Cloud Computing	Edge/Fog Computing
Stores files on systems such as Swift	Stores files on storage devices (i.e. RAM)
Compatible with multiple device types	Only compatible with specific hardware
Connected to the Web and other servers	Physical routing to fog nodes
Gathers data with Ceilometer	Gathers data with physical sensors

Chapter 8 Homework Problems

1 How would cybersecurity affect the development and implementation of the IoT and wearable technology globally?
Cybersecurity affects development and implementation of IoT and wearable technology globally by forcing designers to think about how to protect the devices and implement additional features to maintain a secure and usable product. This may cause slower production of these devices due to additional planning and implementation of security features, especially if there are specific requests for the product such as size, memory and other essential features.

2 Research the commercially available security solutions dedicated for IoT and wearable technology. Compare between their effectiveness based on the targeted area.
A study on ten different IoT-based security systems show that all were very vulnerable. They allowed the use of weak passwords, lacked account lockout mechanisms and proper account information protection so that brute force tactics could be used against them. For example, most of those that involved cameras and motion sensors could give access to the footage to multiple accounts or devices, allowing for potential hackers to easily see the footage for themselves. This shows a lack of confidentiality and authentication/authorization on their end which makes them vulnerable IoT systems.

3 You are working on prototyping a smart garden moisture sensor. Assuming you are still in the early stages of the prototype (i.e. breadboarding), what would you do to secure (a) the software, and (b) the hardware?

A One could secure the software by implementing password or other authentication systems into the device to ensure that only those working on the garden are allowed to change the parameters and specifications of the device.

B One could consider how to add an extra layer of protection from tampering along with protecting the device itself from external damage from water, wind, the earth and other physical factors.

4 Create a table that maps the security goals against threats and attacks mentioned in the chapter.

Security Goal	Threat(s) and Attack(s)
Confidentiality	Spoofing, Information Disclosure, SQL Injection Attacks, Dictionary Attacks
Integrity	Tampering, Repudiation, Elevation of Privilege
Availability	Spoofing, Tampering, Denial of Service, Elevation of Privilege, DDoS Attacks, Physical Attacks/Theft
Authentication/ Authorization	Spoofing, Tampering, Elevation of Privilege, SQL Injection Attacks, Back Door Attacks
Auditing	Denial of Service, SQL Injection Attacks, Brute Force Attacks, DDoS Attacks
Non-Repudiation	Tampering, Information Disclosure

5 You are designing a wearable device that controls a pacemaker. What are the possible threats and attacks? What is your risk assessment strategy?
Some possible threats and attacks include that a hacker could hack into the pacemaker's software and remotely control or tamper with the device to alter the wearer's heart rate to cause harm to the wearer's body. Furthermore, information specific to the wearer could be stored on the pacemaker that could potentially be extracted from the device or modified. The wireless system of the device must be ultimately secured.

6 You are working on a project that involves designing a smart door lock that can be controlled remotely. What are the most important security considerations?
The most important security considerations for a smart door lock include preventing the ability for hackers to either tamper with and unlock the door themselves or to pretend to be the owner of the place where the door is and unlock the door themselves and later hiding the fact that it was unlocked without the owner's knowledge.

7 How would you prevent brute force attacks when planning your IoT projects?
If a password system is used for a device, one could set a limited number of attempts to input a correct password. If the number of attempts exceeds that number, the device either locks itself for a certain period of time and/or lets another device, which would likely be in the possession of some sort of administrator, know of the repeated guessing of passwords, which could then unlock it manually.

8 Research how Mirai and Satori attacks are related. What can you do to prevent such attacks in the future?
Both Mirai and Satori attacks are related by being botnet-based cyberattacks that can attack IoT devices through distributed denial-of-service attacks. Attacks can be prevented by preventing DDoS attacks through re-routing the malicious and spam data and filtering it out before reaching the sites to prevent any major harm from spreading as far as the Mirai and Satori attacks did.

9 Research the different deployment models a DDoS mitigation provider may offer. What would you choose for an IoT-based smart home project?
For an IoT-based smart home project, one would use a cloud-based scrubbing system and a DDoS-aware firewall for DDoS mitigation. Here, the individual devices that make up the smart home could function as they could normally while an outer device in the cloud and the firewall work on filtering out the harmful data from DDoS attacks from reaching any of the devices.
Other potential services could be found at this site:
https://www.f5.com/services/resources/white-papers/the-f5-ddos-protection-reference-architecture

10 Would your soil moisture monitor project benefit from blockchain technology? Why or why not?
While IoT could greatly benefit from the security blockchain could offer, the soil moisture data is not sensitive enough to be worthy of implementing such computationally heavy algorithms.

Printed and bound by CPI Group (UK) Ltd, Croydon, CR0 4YY

16/04/2025

14658581-0002